Simulation
Fundamentals

Prentice Hall International
Series in Systems and Control Engineering

M. J. Grimble, Series Editor

BAGCHI, A., *Optimal Control of Stochastic Systems*
BENNETT, S., *Real-time Computer Control: An introduction*, second edition
BITMEAD, R. R., GEVERS, M. and WERTZ, V., *Adaptive Optimal Control*
BROWN, M. and HARRIS, C., *Neurofuzzy Adaptive Modelling and Control*
BUTLER, H., *Model Reference Adaptive Control: From theory to practice*
COOK, P. A., *Nonlinear Dynamical Systems*, second edition
ISERMANN, R., LACHMANN, K. H. and MATKO, D., *Adaptive Control Systems*
KUCERA, V., *Analysis and Design of Discrete Linear Control Systems*
LUNZE, J., *Feedback Control of Large-Scale Systems*
MARTINS DE CARVALHO, J. L., *Dynamical Systems and Automatic Control*
MATKO, D., ZUPANČIČ, B. and KARBA, R., *Simulation and Modelling of Continuous Systems: A case study approach*
MCLEAN, D., *Automatic Flight Control Systems*
OLSSON, G. and PIANI, G., *Computer Systems for Automation and Control*
ÖZGÜLER, A. B., *Linear Multichannel Control*
PARKS, P. C. and HAHN, V., *Stability Theory*
PATTON, R., CLARK, R. N. and FRANK, P. M. (editors), *Fault Diagnosis in Dynamic Systems*
PETKOV, P. H., CHRISTOV, N. D. and KONSTANTINOV, M. M., *Computational Methods for Linear Control Systems*
ROGERS, E. and LI, Y., *Parallel Processing in a Control Systems Environment*
SÖDERSTRÖM, T. D., *Discrete-time Stochastic Systems*
SÖDERSTRÖM, T. D. and STOICA, P., *System Identification*
SOETERBOEK, A. R. M., *Predictive Control: A unified approach*
STOORVOGEL, A., *The H^∞ Control Problem*
WATANABE, K., *Adaptive Estimation and Control*
WILLIAMSON, D., *Digital Control and Instrumentation*

Simulation Fundamentals

B. S. Bennett

Prentice Hall

London   New York   Toronto   Sydney   Tokyo   Singapore
Madrid   Mexico City   Munich

First published 1995 by
Prentice Hall International (UK) Limited
Campus 400, Maylands Avenue
Hemel Hempstead
Hertfordshire, HP2 7EZ
A division of
Simon & Schuster International Group

Typeset in 10/12pt Times
by Keyset Composition

Printed and bound in Great Britain by
Bookcraft, Midsomer Norton

Library of Congress Cataloging-in-Publication Data

Bennett, B. S.
    Simulation fundamentals / B.S. Bennett.
        p.   cm. – (Prentice-Hall International series in systems and control
engineering)
    Includes bibliographical references and index.
    ISBN 0-13-813262-3 (cased)
    1. Computer simulation.   2. Discrete-time systems.   I. Title.   II. Series.
QA76.9.C65.B46   1995
003′.35365–dc20   94-37138
                                                                             CIP

British Library Cataloguing in Publication Data

A catalogue record for this book is available from
the British Library

ISBN 0-13-813262-3

---

1   2   3   4   5   99   98   97   96   95

For my wife Ann
and my daughter Heidi

Contents

Preface

This book is intended to introduce the novice reader to the world of simulation, to illustrate the part that simulation can play in many different situations, and to provide some insight into the implementation of simulation, using typical personal computer hardware and software available today. In these respects too, the book is intended to be of use to people with some simulation experience.

As the novice reader will discover, there are two very different techniques given the name of simulation. The first of these to be introduced is *discrete simulation*. This is very much a decision-support tool of management scientists, inventory control experts and those people interested in the efficient operation of queuing systems, where waiting times have to be minimized and utilization of expensive facilities maximized. This mode of simulation may be more familiar to the reader in the guise of *operational research*, of which it is one of the available tools for solving management problems.

This book lays more stress on the second technique, *continuous simulation*. This mode of simulation is concerned with the dynamical behaviour usually, but not always, of engineering systems, as described in terms of mathematical relationships. Engineers and scientists, even if they have never applied simulation techniques before, are certainly familiar with the mathematical relationships that they use to describe system behaviour. Workers in many other disciplines, for example medicine and economics, may find it useful to apply continuous simulation to their problems by invoking similar modelling methods. On a lighter note, most addicts of arcade games may not realize that the virtual reality of the racetrack or air combat is an example of continuous simulation at work.

Readers are assumed to have some knowledge of mathematics, perhaps to first-year degree level, and therefore to be acquainted with calculus, algebra (including linear algebra) and statistics. Some familiarity with programming languages – particularly FORTRAN – would be useful, but is not essential, since all program listings given are commented. For the same reason, no knowledge of simulation languages is required. If the book is to be used as a coursework text, a single academic year should be sufficient to cover the work. For use in an industrial environment, readers can simply refer to the sections relevant to their needs.

There is little concern with the modelling aspects of dynamical systems. The main purpose is to assist the reader to develop a working simulation, starting from the point where a set of mathematical relationships – the simulation model – is available. There are many implementation problems to be faced in all but the simplest cases, and it is the aim of the book to show how such problems may be overcome.

Although the book is mainly about continuous simulation, some insight is given into how combinations of the two main techniques can improve the quality of the simulations obtained. This applies whichever way the modelling emphasis lies.

As regards simulation hardware, almost total emphasis is placed on the use of personal computers. Examples have been set up and program listings given for software packages designed to work on IBM PC-type computers and 100% compatible clones. This hardware medium was chosen because of its ubiquitous nature, and consequently the greater likelihood that the interested reader will have access to at least one of the two packages used to run the programs supplied.

In making a choice of software media under which to run the simulation examples, two criteria had to be satisfied. On one hand, it was considered very important to give the reader an introduction to one of the more popular continuous simulation software packages. On the other hand, it was felt that the use of an interpretive medium would be of value in working through tutorial examples, to enable quick changes, optional output, and illustration of the mechanics of simulation. This is not readily done from a compiled program without extensive access to debugging facilities, which are compiler dependent.

In order to illustrate the development of a structure suitable for continuous simulation, which includes useful elements of discrete operation, a set of routines was written to function as a simple simulation package under MATLAB^{TM}. The facilities provided by MATLAB, together with its interpretive nature, proved useful for tutorial examples.

The Advanced Continuous Simulation Language (ACSL^{TM}) was chosen to illustrate the use of a general-purpose simulation package. This is available for IBM PC-compatible machines, and has all the features desirable in combined (continuous-plus-discrete) simulation. It is one of the more widely used commercially available packages in use today, of which the PC version is but one of many computer implementations.

The contents of the book have been distilled from many years of simulation experience in the electrical power and chemical process industries, and from a considerable period teaching the subject at university level. In consequence, it was hoped to ensure its suitability both for coursework and for use in industry.

I should like finally to acknowledge with much gratitude the help given to me by Mitchell and Gauthier Associates, in connection with the implementation of ACSL/PC, and the close links I have had with them over the years in running ACSL on computers and workstations. I should like to acknowledge the help given by The MathWorks Inc. in respect of the MATLAB routines, and the use of SIMULINK. I should also like to thank Messrs Rapid Data Ltd, and in particular Bill Havranek for his assistance in connection with ACSL and SIMULINK.

B. S. Bennett, 1994

TRADEMARK ACKNOWLEDGEMENTS

ACSL, ACSL/PC, and **ACSL/Graphic Modeller** are trademarks of Mitchell and Gauthier Associates Inc.
MATLAB and **SIMULINK** are trademarks of The MathWorks Inc.
Microsoft and WINDOWS are trademarks of Microsoft Corporation.

1

Systems and Models

1.1 WHAT IS SIMULATION?

Simulation as a technique or a set of techniques was probably unknown to most people before the late sixties when the space race and, in particular, the race to get to the moon, pushed it into prominence. Before that time, the only people who knew much about simulation were those who actually used it to solve problems. However, when the first landings took place on the moon, people noticed how it appeared possible to predict with considerable accuracy just what was going to happen at the next stage of the lunar module's journey. For example, it was possible to predict splashdown in the Pacific Ocean to within a few miles when the spacecraft was still a long way off. Also one could observe how the astronauts were given course corrections to ensure that the lunar module achieved its intermediate goals at each stage of the journey. In this way, simulation played a very important part in a drama that was witnessed by millions of people the world over. This appeared to be the time when the term simulation really came into everyday use. However, that is not the same as understanding what simulation is all about. Someone new to the subject might ask: What does simulation really mean? Why and how is it used?

It may be a coincidence, but also in the late sixties, a spate of definitions of simulation appeared in all the relevant literature. Many of the eminent simulationists of the time (McLeod, 1968, Gordon, 1969) gave their definitions, within which many common features emerged. Simulation was variously described as an 'art' whereby one could develop models to represent real or hypothetical systems. Simulation was also acknowledged to be a technique or a set of techniques whereby the development of models helps one to understand the behaviour of a system, real or hypothetical.

1.1.1 What is a System?

In order to complement the various definitions of simulation, it was necessary to try to establish just what constituted a system. From the different interpretations that arose (Gordon, 1969, Shearer, Murphy and Richardson, 1967), a system could be considered to be a collection of objects, parts, components, call them what you will, which interact with each other, within some notional boundary, to produce a particular pattern of behaviour. The idea of a boundary was necessary to separate the system from the rest of the universe, to keep the task of studying its behaviour within reasonable limits!

1.1.2 What is a Model?

The logical development as far as simulation is concerned is to establish what constitutes a model, since the 'art' of simulation is essentially the modelling of systems, with the use of computers. In order to answer that question, we have to consider, very carefully, just what information we expect to obtain from the manipulation of a model during simulation studies. Once we have established that, even in general terms, then we can begin to decide what form our model should take.

1.2 HIERARCHY OF MODELS

In the broadest sense, models take two forms: physical (iconic or replica) models, or abstract (notional or mathematical) models. It is the abstract model that is more relevant to the idea of general-purpose simulation (that is, simulation from which it is expected to obtain much more than a narrow range of information about the system). However, it is useful in the wider context for us to consider and contrast both basic forms to see why this should be so. Figure 1.1 shows a simple tree structure displaying a hierarchy of model structure as used in this book. The heavier branch lines connecting model attributes indicate the degree of importance attached to them.

1.2.1 Physical (Replica Models)

In the course of development work, many industries require the construction and testing of replica models for specific purposes. It used to be the case that aircraft were the only form of transport subjected to wind tunnel testing during the course of design. This was to try to achieve an aerodynamic optimum in terms of low drag coefficients. Reducing drag reduced fuel consumption, which gave a better payload, and also probably reduced the required engine thrust needed to achieve a given airspeed. In other words, the economics of flight necessitated wind tunnel testing of replica models as part of the airframe design process.

  In recent years, with motoring costs rising (fuel prices being a major

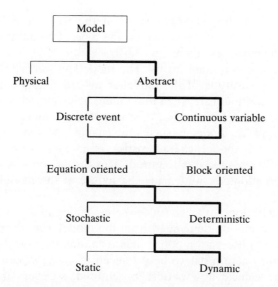

Figure 1.1   Hierarchy of model structure.

contributor), the motor manufacturing industries too have found it expedient to subject replicas of their car body frame designs to wind tunnel testing, again to try to minimize drag factors. That this has been successful is reported in the trade literature in ever reducing drag figures. The negative side is the tendency for car shapes to become very similar!

Chemical process industries may need to construct so-called *pilot plant* in order to carry out design testing for a new process. The data obtained from the small-scale version can then be scaled up to use in the design as required.

In our world of simulation, physical modelling plays no part at all, and will not be considered further. One good reason for this is the narrow range of interest served by physical modelling. Even in pilot plant testing, the aims of the exercise will seldom extend over the whole process. It is quite likely that the studies carried out would yield information designed to provide a quite limited range of answers. General-purpose simulation should be able to do more than that, and that is where the abstract model comes in.

1.2.2 Abstract (Mathematical Models)

A notional or indeed mathematical model of a system is an abstraction of the reality it is meant to portray. By putting the modeller's ideas of the real system behaviour into words, mathematical relationships or any other abstract form, an abstract model can take shape. It is quite clear that the ideas in the modeller's mind concerning his or her perceptions about the way the real system should behave, will have considerable influence on the type of model that evolves.

For simulation purposes, therefore, abstract modelling is all important and generally takes one of two main forms. The modelling feature that brings about this dichotomy is the attitude of the modeller toward simulation time. If model behaviour changes continuously, such that model behaviour characteristics should be accessible at every instant of time, then a *continuous* model results. If, on the other hand, the type of model is such that changes of state occur only at set instants of time, the model characteristics remaining constant at points in between, then a *discrete* model is required. This distinction is an example of the type of abstraction adopted, influencing the nature of the model that eventually takes shape.

By way of illustration, let us consider the modelling of a vehicular traffic flow situation. If the information demanded from the model is an average vehicle flow rate past a particular observation point, then a continuous model may be needed, especially if the information has to be produced in an unbroken form. Constant observation of the vehicle flow pattern is required, so every instant of time is equally important. If, however, it is specified that the model should give information about the individual arrival pattern of vehicles at the same observation point, then a discrete model may be more suitable. The only changes to the state of the model occur with each arrival at the observation point. Time is of no consequence in between each arrival. In practice, there is often a degree of overlap between the two model types, because of the nature of the real system behaviour and the consequent information required from the model thereof. The modelling mix may be in any proportion deemed by the modeller to suit the task in hand.

1.3 MODEL CHARACTERISTICS

The model types outlined above are not the only contrasting features we have to consider in the course of development of a simulation model. Early simulation work was almost all carried out for engineering system design, using analog computers to solve the mathematical relationships describing the expected patterns of model behaviour. The very nature of an analog computer requires the model formulation to be in a *block-oriented* form, to correspond with the analog components. Modelling characterized by functional blocks with input/output interconnections was adopted when digital processors were first used for simulation work. This was done to attract dyed-in-the-wool analog computer programmers away from their machines towards the new digital simulation software. Indeed, block- or function-oriented modelling is still in use today in most of the well-known simulation software packages.

The disadvantage of a purely block-based modelling approach is the need for very many blocks to represent the complexity present in a typical single-line mathematical relationship. Also departure from the form of the relationship can, for many, be accompanied by a loss of meaning upon dissection into many function blocks. The *equation-oriented* modelling form, facilitated by the use of modern high-level computer programming languages, preserves most of the identities of the mathematical relationships embodied in the procedural coding statements, and quite complex modelling can be present in one line of code. Most general-purpose simulation software is procedural and therefore equation oriented, but elements of block-based modelling are used, for example, for the representation of arbitrary functions, and special non-linearities.

If we consider the different systems we may be called upon to model, we would very soon come across uncertainty in various forms. We would most probably be uncertain about different aspects of the model structure itself, being unsure whether a particular relationship was the best way to describe some model feature. Even if we were sure about the model construction details, we would be uncertain about the authenticity, integrity and values of some of the data to be used in implementing simulation studies. In the simulation of engineering systems, we usually take a decision to ignore the uncertainties of model and data identification, and to treat the models as being *deterministic*. If the uncertainties are of little importance compared with the general dynamical behaviour of the model, then such decisions are valid. If the uncertainties are sufficiently important to affect model dynamics, then they cannot be ignored.

When we consider the behaviour of discrete systems and the corresponding models, a different picture emerges. Discrete modelling tends to involve the interaction of people with systems. Continuous models usually focus on the behaviour of engineering systems, to the exclusion of the human factor. People inject a great deal of uncertainty into the behaviour of any system with which they come into contact. They respond differently to a situation, react at different rates to an incident, and have a variety of ways in which they may interact with a system. So-called man-in-loop or human factor involvement in engineering systems requires careful modelling to obtain the correct mix of *deterministic* and *stochastic* effects. However, the greatest uncertainty is to be found in the discrete modelling of queuing situations where arrivals may be random, processing times can be variable, and the consequences for the utilization of resources will be unpredictable. In this case, *stochastic* or *probabilistic* modelling elements have to be included at every point where variability needs to be taken into account.

Among the many interpretations of the word *simulation* is its extension to cover the modelling of systems or processes that are in steady state. This is particularly the case in the chemical process industries, where steady-state or *static* modelling is important for flowsheet development. The reality of the situation is that dynamical system modelling includes static relationships as a special case, which may be extracted from the modelling process if required. In this book we are concerned wholly with the development and implementation of dynamical model simulations. Static processes as such will have no part in our deliberations.

The above outline of model characteristics shows how, in terms of the attributes we have covered, a model may well consist of mixes of each of the contrasting pairs. There appears to be no hard and fast rule which says that a continuous model shall be wholly deterministic, or that a discrete model shall be block oriented. For the most part, however, continuous simulation will be treated as deterministic, and a discrete model will be stochastic. There are several other model characteristics, not discussed above in the model hierarchy, but which are recognizable as being distinguishing features of either continuous or discrete models.

Dynamical models which feature sets of *differential equations*, either ordinary or partial, or indeed both kinds, will almost certainly be continuous models. The solution of differential equations does not come into discrete simulation methodology. Although a model may consist of a discrete part interacting with a continuous part in a *combined simulation*, it is the individual parts considered separately, for which we seek these distinguishing features.

In the same way, we may look for the presence of sets of *algebraic equations*. Again a continuous model will probably contain many to define auxiliary system variables, whereas in discrete simulation there may be some algebraic relationships, but they will be few in relation to the size of the model.

If we look for the presence of what we may call *logical relationships*, or in other words, decision-making processes, we will find that in discrete models, decision-making is a very important feature, and the model operation is driven by the interaction of many such processes. Continuous models, on the other hand, may exhibit very little in the way of decision-making to alter the course of events as a simulation proceeds.

It is worth stressing again the significance of time, as mentioned in Section 1.1.2, since the treatment of time advance is the most important distinguishing feature of models, and has a great bearing on the ways in which models are mechanized on the different computational devices available.

1.4 HOW TO USE THIS BOOK

The book is intended to be a primer for the application of continuous simulation for those of you who are new to the subject. You should read the first two chapters in order to appreciate the scope of simulation as a whole and to sample the various flavours of simulation before getting down to some detail. Although the book is predominantly about continuous simulation, Chapter 3 acknowledges the importance of discrete simulation by presenting a broad outline. This should be sufficient to cover the application of discrete concepts in the remainder of the book, but for those who become interested in this aspect, pointers will be given to books that cover the subject in more detail. Chapter 3 makes essential reading for both beginners and those of you with simulation experience. Proper application of event-processing techniques in later chapters depends on a good basic understanding of discrete concepts.

If you are a beginner, then you should not miss out Chapter 4, simply because the hardware described may seem obsolete. An understanding of the basic concepts of analog computation will lay a good foundation for you to be able to solve some of the problems that beset even experienced simulationists. Although the demise of the hybrid computer is mourned by few on account of the many programming and operating difficulties, this hardware, too, has much to teach the simulationist about computer control and related simulation problems.

Chapters 5 to 8 should be regarded as core material for all readers, regardless of level of competence. Chapter 5 follows on naturally from Chapter 4 in that analog computing concepts are carried straight over to the digital processing scene. The better attributes of analog computation are set as targets for digital simulation to reach, and digital computation's own implementation problems are contrasted with analog hardware programming difficulties.

Numerical integration is a problem area not associated with analog computing machinery, but it is one of the key issues with digital computation. Chapter 6 looks at the different techniques of numerical integration and makes suggestions about the best methods to consider for different kinds of model implementation.

Another difficulty associated with numerical integration is the occurrence of discontinuities in a continuous simulation. Chapter 7 is devoted to this subject, and lays the foundation for the treatment of discontinuities by event-processing methods. The application of this treatment to the simulation of sampled-data systems is demonstrated in Chapter 8. This is of importance for readers wishing to carry out simulations of computer control problems. Some of the lessons learned from the hybrid computer will be evident in this treatment.

The material in Chapter 9 is meant for those of you who may wish to apply general-purpose simulation methods to the solution of partial differential equations. If we keep the theory to a minimum, it is possible to present the model manipulation developments in an intuitive way, leading eventually to working simulations.

Chapters 10 and 11 present case studies taken from industrial simulations. Although the examples are somewhat simplified versions of the 'real thing', the principles and steps taken to implement the working simulations have, however, been retained, down to the last detail.

The final chapter in the book is a 'look ahead' to what we may see as being some of the likely developments in the simulation scene, in hardware, software and techniques. Although speculative, this does point to the way developments are moving at present in what is a very exciting and interesting area of work.

Besides the case studies, modelling and simulation examples have been presented at various points throughout the book. Some of these are for illustration purposes only, but most are intended for you, the reader, to try out for yourself and 'play around' with in time-honoured fashion.

Models are presented for implementation by means of two media for digital simulation. The first is *ACSL* which is a world leader in terms of general-purpose simulation software. The latest versions make good use of the graphical user interface environment. The second is the *MATLAB* computational software package and its associated programming language. A set of tutorial routines has been developed for use with this book to enable simulation exercises to be carried out on a variety of models. Although not as comprehensive as the likes of ACSL for addressing all aspects of continuous system modelling and simulation, the routines (given the collective name of *MATSIM*) nevertheless have several enhancements over and above straightforward numerical solution of differential equations. They will carry out run-time plotting, iterative solution and event processing, and, being written in the MATLAB language, it is possible for you, the reader, to examine in detail most aspects of the way in which a digital continuous system simulation package might be expected to work.

1.5 REFERENCES

ACSL (Advanced Continuous Simulation Language) (1991) Reference Manual, Edition 10.0, Concord, MA, USA, Mitchell and Gauthier Associates.

MATLAB (1992) Reference Guide, Natick, MA, USA, The MathWorks Inc.

GORDON, G (1969) *System simulation*, Englewood Cliffs, NJ: Prentice Hall.

MCLEOD, J (1968) *Simulation: The modelling of ideas and systems with computers*, New York: McGraw-Hill.

SHEARER, J L, MURPHY, A T and RICHARDSON, H H (1967) *Introduction to system dynamics*, Reading, MA: Addison-Wesley.

2

Overview of Simulation

2.1 WHAT IS SIMULATION? WHY DO WE NEED IT?

From the material presented in the first chapter, you, the reader, will have gathered that simulation is something along the lines of a technique or a set of techniques for examining the dynamical behaviour of (abstract) models. You should also have noted that this is done by implementation on some computational device. By manipulating and testing the model in various ways to ascertain the model behaviour, we would hope to be able to ascribe that behaviour to the real system that we are trying to emulate. It does not matter if the system does or does not exist. Either way, the application of simulation to the model of the system should, in theory, be quite feasible if the modelling relationships are known or can be inferred.

2.1.1 Simulation's Role in Engineering Design

It is one of the great strengths of simulation that the system upon which a simulation model is based need not exist. This enables simulation to be applied to engineering design from the outset. Industries which use simulation a great deal have come around to this way of thinking, and the vogue is now to regard simulation as one of the many design tools that will be applied to the engineering system right from the conceptual stages. The advantages of this approach are obvious. The information yielded by simulation testing of dynamical models will be very different from that obtainable from other design tools, and simulation is the only technique to allow an examination of the dynamical behaviour of system models. It is therefore capable of giving insight into aspects of transient misbehaviour, such as temporary excursions outside defined constraints, that is not available from the use of other design tools. Simulation has an important role in engineering design.

2.1.2 Testing for Safety

The role of simulation in testing for safety is obviously one of the facets of application to engineering design. Any information that has a bearing on the future safe operation of an engineering system has to have considerable value for design purposes. The value of simulation in this context does not only apply to hypothetical systems. It can be applied just as readily to existing systems, for which the application of safety tests would be a hazardous venture in real life. For example, it would be unthinkable to create a maximum credible accident situation in a nuclear power station in order to test whether the safety systems would all operate in a satisfactory manner to bring about a fail-safe condition for the reactors. Accident testing of this nature has to take place by means of simulation studies, and the results interpreted in the light of one's faith in the models used. It is not only in the area of accident studies that simulation comes into its own.

2.1.3 Option Testing

In the conceptual stages of design, many different options will be postulated for the engineering system under consideration. Simulation will assist in arriving at the 'best' design options at all stages throughout the design process. The option choice will be widest at the outset, and will narrow down as the design becomes more 'firm', but there will be opportunities for choice right to the end. An example of this latter effect concerns the question of the implementation of control for engineering systems. This raises its own set of problems which embrace all three of these topics. The design of control systems has implications for safety, and testing of the various design options is best done by means of simulation. The design of control systems is often left until the latter stages of the overall system design, something that is arguably not good policy. Many engineering systems that appear robust in most aspects of their performance may be difficult or impossible to control in practice, even with the best control implementations. Some attention to the model dynamics at an early stage may help to alleviate such problems.

Although the above rationale for simulation has taken on the flavour of engineering and hence continuous simulation, the same precepts apply to the world of discrete simulation. It is most unlikely that a stock control manager of a warehousing facility would allow anyone physically to meddle with the organization and methods whereby stocks were maintained at 'optimum' levels. Improvements to the system based on some criteria (often operating costs) would have to be justified on the basis of simulation models. Although safety is not necessarily a criterion in such instances, design economics certainly plays an important part, as is the case with engineering systems.

2.2 A SHORT HISTORY OF SIMULATION PRACTICE

Continuing on the theme that simulation is the study of abstract models of systems, using computers to mechanize the model implementations, it is necessary for us to consider what computers are best for the purpose and why. It may seem rather odd to pose such a question in the light of two facts about simulation:

- The first continuous simulation was carried out through the medium of analog computation, simply because that was the only way of doing it at the time. Digital computers, as we know them, were not generally available until the early sixties.
- At present, almost all continuous simulation is implemented on digital processors.

Why then has the transformation taken place whereby analog computers, at one time the only general-purpose simulation media in existence, have all but disappeared from the simulation laboratory? Does it mean that the analog computer was such an appalling device to use for simulation, that it was bound to be scrapped in favour of an alternative when this was available? Was the digital processor seen to be a superior simulation tool at the outset? The answers probably lie somewhere in the middle of these extremes. Early analog computers were notoriously unreliable machines, particularly so when you consider that the electronics was based on now very old-fashioned valve (vacuum-tube) technology. Maintaining hundreds of DC amplifiers in working order, for the implementation of even a medium-sized model, was a full-time job, and one could never be absolutely sure about the results. The larger the model, the greater the number of analog components required, with a corresponding increase in overall unreliability.

Early digital machines were not much better for the same reason, coupled with the fact that programming was a nightmare. The advent of transistor technology, followed by integrated circuitry, meant that both digital and analog computer hardware became much more reliable. The increase in solution speed obtained with every advance in digital hardware technology whittled away the speed advantage hitherto enjoyed by the analog computer. The one great redeeming feature of the analog computer as far as continuous simulation was concerned, was the ability to carry out the operation of integration very quickly indeed, and at a rate independent of the size of the model. This meant that differential equations could be solved at very high speed, useful for carrying out many hundreds of simulation runs in a short space of time. Advances in integrated circuit technology, however, meant that ever-increasing solution speeds could be obtained from digital processors, allowing very large problems to be solved. Parallel operation and the technology of transputers have contributed to this achievement.

Apart from the hardware advances, ease of programming or implementing a model on the relevant computer hardware had to be taken into consideration. Analog computers were never the easiest devices to program. Problem preparation in the way of scaling and 'patching' could lead to many implementation errors that were not always easy to trace. Digital processors, on the other hand, with high-level programming languages, good operating systems, and helpful utilities

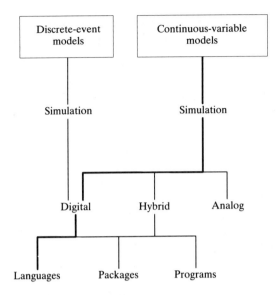

Figure 2.1   Hierarchy of simulation.

for program debugging, became relatively easy to program, and setting up a simulation model developed into a somewhat easier task on a digital machine.

The hybrid computer enjoyed a relatively brief period of popularity during the 'great space race', but has since almost vanished into oblivion. The rationale behind the development of the hybrid was the desire to have the 'best of both worlds' in combining analog and digital computers together into one machine, each communicating with the other through an 'interface'. Several advantages were gained from this combination, most of all in the use of digital software to assist in programming and setting up the analog machine. However, the complexities of model implementation and debugging brought about the demise of the hybrid in favour of a wholly digital approach.

As far as discrete simulation was concerned, there was never a conflict of interests for hardware implementation. Operational research as a management science tool only developed when digital computer hardware was well advanced and good-quality mainframe machines were available. This historical timing, together with the fact that discrete models are best suited to digital implementation, meant that analog machinery has hardly ever been used.

It looks, therefore, as if the digital processor is going to reign supreme as the computational device of the future. This certainly appears to be the case, for all present-day advances are in this field, in terms of both hardware and software. Figure 2.1 shows a hierarchy of simulation, starting from the two main kinds of model. We may progress downward through the layer showing the hardware used to implement these model types, ending up at the forms of software used on different machines.

2.3 SIMULATION SOFTWARE: PROGRAMS, PACKAGES OR LANGUAGES?

For the implementation of a simulation model on a digital or a hybrid computer, some software development is necessary. If we look at software under these three headings, this should be sufficient for us to cover all possibilities. The earliest digital computer attempts at model implementation probably took place within an academic environment, where researchers sat down and wrote one-off special-purpose programs to carry out specific modelling and simulation tasks. Such programs had little value for any other similar work, apart from the experience gained in development. A natural outcome of this was the evolution of general-purpose programs for continuous simulation. In each case, the modules necessary for carrying out all the tasks associated with simulation were made available within a single software package. Many such packages have appeared over the years, usually from academic departments, but also not infrequently within industrial environments. The essential features of such packages are:

- at least one numerical integration routine
- a model description input module (usually a procedure)
- a data input module
- an output module for the simulation results
- a graphical facility to display model simulation transients.

All these features, and more, were combined in a single package written in a high-level programming language such as FORTRAN. A user of such a package needed to have a knowledge of FORTRAN in order to write the model description, and the model variables had to conform to programming conventions. The main advantage of using such general-purpose program packages lay in the ease of debugging through direct access to compiler-dependent debugging utilities. Not only was it relatively easy to trace model implementation problems, but if the whole package were written in FORTRAN, and the source code available, debugging could be extended to other areas, for instance the numerical integration routines. If you, the reader, do not mind programming, then writing your own simulation package is a very good way of learning all about simulation.

In the earlier days of mainframe digital computers, the manufacturers who were obviously out to increase sales of their products knew that a sound software base would assist greatly in this respect. Simulation by means of analog computers had been an area of engineering activity which appeared to be financed by a bottomless purse, when it came to defence interests, and the digital computer manufacturers considered that they too should have a 'piece of the action'. Thus began the development of what became known as *simulation languages*, which at first were simply analog computer emulators with all the salient features embodied in digital software packages. The block structure of such implementations was soon seen as a hindrance, especially when the number of such blocks was severely limited by what could be displayed on the mimic analog patching diagrams of the day. However, the nature of the block structure virtually divorced the simulation syntax (rules of operation) from that of the FORTRAN or other high-level language compiler in use. This was seen as an advantage by engineers who did not at that time want to have anything to do with programming! As a result, a more procedural form of simulation syntax evolved, again virtually free from the restrictions of FORTRAN. The various implementations became true simulation languages and ceased to be emulators of analog computer operation. Most commercially available packages in use today belong to this class of software, the development of which has advanced a long way from those early versions.

Although simulation languages are free from most compiler restrictions, they do have a syntax of their own. This is designed, in general, to ensure that when the model description is passed through a translator module, the resulting high-level language code is error free and virtually guaranteed to compile correctly. Most simulation languages translate their own code into FORTRAN, and the model description appears as one or more FORTRAN subroutines or procedures which are compiled in the normal way. If you have never encountered digital simulation before, then in spite of what the computer scientists may wish upon you, you could find it advantageous to learn FORTRAN if you have not already done so. Historically, most simulation packages had their foundations in FORTRAN, and with the huge investment already made in software in that direction, it is highly unlikely that changes will take place in a hurry!

In terms of the software used for simulations of all kinds at the present time, simulation languages hold sway over everything else. It is worth while for us to consider why this should be so. All simulation languages are available commercially, with different companies developing and marketing their products for a living. Most users of simulation software, in particular large corporations, do not generally want to be concerned with the development and maintenance of software packages. They are usually quite prepared to pay an annual fee for the privilege of having access to regular updates of the software, and to hot-line help in dealing with problems. In return, the software marketing companies work to enhance their products to keep or increase market share, and will, in the process, usually be keen to listen to their users' comments. The process seems to work quite well, but there are disadvantages. Users with problems may have to wait some time for assistance, and the proprietary nature of the software packages usually means that program source listings are not available to users. If problems concern the model alone, this does not matter, but if the simulation software is at fault, the effect can be devastating for the user. At this stage, we can note one important fact about simulation. The more remote the package is from the end user, and the longer the cycle time between starting a simulation and getting the results, the more difficult it is for the simulationist trying to develop a 'feel' for a model and the way it should behave. In this one respect above anything else, the analog computer scored hands down in its day.

2.4 SIMULATION STRUCTURE

Although you may not think so, there is a definite 'structure' to the way in which a simulation is carried out. If you are not familiar with this concept, perhaps some examples from everyday life will illustrate the point. It is possible to do this from the points of view of both discrete and continuous modelling.

2.4.1 The Game of MONOPOLY™ – a discrete model

Most of you will have come across the game of MONOPOLY at some time in your lives. It has to be one of the most popular board games in existence, and serves as a good everyday example of discrete simulation. Let us examine the procedures followed during the course of an average game.

The first stage is setting up the game: someone is forced to be the 'banker', arguments take place concerning the allocation of the play tokens, an initial sum of money is dealt to each player, and the game is ready to begin. All this constitutes the *initialization* phase of this 'simulation'.

The second stage is where it 'all happens'. MONOPOLY is a simple simulation of the property market, and works generally in the same way as a discrete simulation. One feature of such simulation is the action of a random-number generator to create the variability and consequent stochastic behaviour mentioned in Section 1.3. The random-number generator used in a game of MONOPOLY is the pair of dice supplied. Throwing the dice generates 'random' numbers and galvanizes the players into activity. Simulation time advance is effected by movement of the tokens around the board. Events are generated by landing on a particular square. Events are processed by carrying out mandatory instructions (if so required) or otherwise by taking decisions concerning the purchase of title deeds, or houses and hotels. This feverish activity constitutes the *dynamic* phase of the simulation. It has a habit of going on rather a long time, as fortunes change, and players stave off insolvency by fair means or foul. Eventually, however, the game will come to an end when one player has forced all others into bankruptcy. A different procedure may be followed if a shorter game is desired. It may be agreed to play for a fixed length of time, at the end of which the richest player is declared the winner.

Whichever play procedure is adopted, the final stage of the game takes place when play ends. This consists of working out the wealth of the winner at the end of a long game, or doing all the sums to determine the winner of the fixed-time contest. The final 'post mortem' on the state of play is the *termination* phase of the game.

2.4.2 Arcade Games – continuous models?

Continuous simulation has its counterpart in the world of games and amusements. Arcade games (the ones you played in amusement arcades when you should have been at school) are everyday examples of continuous simulation. The claim to be true examples of simulation may in fact be enhanced by the ready availability of so many of these games in the form of software for personal computers or dedicated games computers.

It is easy enough to visualize the procedure needed to play one of these games as consisting of three stages, similar to those in a game of MONOPOLY. The *initialization* phase for an arcade game consists of inserting the fee required to activate the machinery. The equivalent action on a personal or games computer involves setting up the game to user requirements, with or without noise effects, etc., as desired.

The *dynamic* phase is the highly interactive involvement with the game through the medium of controls appropriate to the action required. Air combat activity will probably necessitate the use of a joystick for manoeuvring the simulated 'aircraft' with a push-button on the top to 'fire' the weapons. The duration of this phase depends upon the location of the game. In an amusement arcade, the fixed time given for the fee paid causes the activity to cease, while on a personal computer, the onset of boredom is the usual terminator, unless the game itself is time limited.

The *termination* phase itself does not require much action on the part of the player. Some games keep a tally of highest scores, and giving the present player the information is about all that happens, apart from suggesting that the player might like another game!

2.4.3 Simulation Structure in General

The simple examples of everyday simulation given above suggest that there is indeed a course of action that is followed every time a simulation is carried out. This procedure has a definite structure to it that is common to all kinds of simulation: continuous, discrete, or combinations of the two. It does not matter either what form of computer hardware is in use for the simulation, the general procedure is the same. Put very simply, the whole operation of carrying out a simulation run has three separate stages which are outlined below:

- *Initialization*   The model is set up on the computer. This means that the abstract model defined by the simulationist is manipulated into a form which will enable it to be coded in a digital computer high-level language, or a simulation language. If an analog computer implementation is planned, the physical interconnections between the analog blocks describing the model are made on a 'patch panel'. The problem implementation should then be checked out statically and dynamically to ensure that all is well before commencing a simulation run. The initialization stage is then concluded by the input of parameter data (values which may change from one run to the next). This data will include the initial conditions for the simulation.

- *Dynamic stage*　The model is 'set in motion' whereby it is allowed to exhibit its dynamical behaviour under the action of the simulation time advance mechanism. A simulation run may be programmed to proceed for a fixed period of simulation time, or to terminate upon the fulfilment of certain conditions notified to the model in advance.
- *Termination*　When the run is complete, either normally or abnormally, the task can begin of carrying out a 'post mortem' of all that has happened. This may be pre-programmed to follow a particular course under the action of extra software incorporated to carry out a specific analysis of the results. On the other hand, it may simply involve the simulationist examining the results obtained and making decisions as to the next course of action. If a further run is required, new parameter values will be set up and the simulation will return to the initialization stage. Otherwise, further work will be carried out to obtain all the information needed from the run to include in some form of presentation of the results.

As simulation advanced toward the zenith of the 'space race', a recommended specification was published (Strauss, 1967) for a Continuous System Simulation Language (CSSL). This specification listed all the features which had been considered desirable in a simulation language for continuous simulation. One of the salient parts of the report was the specification of simulation structure. This structure centred around the three-stage process outlined above, and it remains in use to this day.

2.5　METHODICAL APPROACH TO A SIMULATION PROJECT

The above structure outline applies to the execution of a simulation from the point of implementation of the model on some computational device through to the point at which the simulation results are obtained. Nothing has been mentioned about the process of developing the model and the iterative testing and debugging that must take place before a simulationist is satisfied with the outcome of the model development. The horizon for model development and testing is far

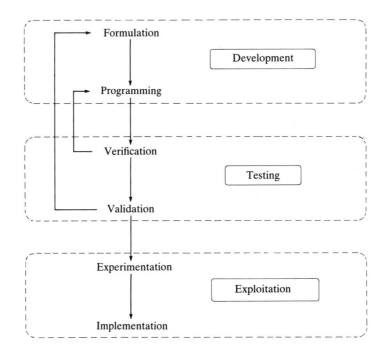

Figure 2.2   Stages of a simulation project.

larger than that of the straightforward simulation structure. It is worth giving some consideration to the steps needed for a successful conclusion to this larger exercise. Figure 2.2 shows a flow diagram of the steps involved in the implementation of an overall simulation project which may necessitate the execution of many hundreds of simulation runs. With reference to this figure, we may consider these steps one at a time.

2.5.1 Model Development

If you are going to implement your abstract model on some computational device, you have to start with an idea of what you expect from the behaviour of the real world system you are modelling. This will be *formulated* in a manner that can be handled by the computer you are using, and constitutes your hypothesis about the way that system should function.

Having formulated your model, you will need to define it in terms of a computer program of some sort. The term *program* is used loosely to describe the communication that has to be created between you and the machine, in the form either of coded procedural statements, or instructions to set up an analog machine. The implementation of the program gives you a starting point for a series of tests with your model.

2.5.2 Model Testing

When you have managed to get your model implementation converted into a working simulation on your computer, the next step is to check the behaviour of the model under specified conditions, to ascertain that it is behaving in the manner that you expect. In other words, you are checking to see whether or not your model implementation is a correct embodiment of the hypothesis you set out at the beginning. In a sense, you are 'debugging' the model at this stage, through a process of *verification*. It does not matter yet whether or not the model represents the real system. It is sufficient for you to know that it is behaving as you expect it to do.

Your simulation is now working to your satisfaction, and the model is behaving in the manner you consider to be representative of the real system. It is very tempting for simulationists to think that just because a simulation is working in what appears to be a satisfactory manner, it must be yielding the correct results; in other words, it emulates closely the real system behaviour. Unhappily, this is seldom the case, unless the simulationist concerned is lucky enough to know the details of the system intimately, and is expert at implementing these in a very good model.

In most cases, it will be necessary to carry out a series of checks to determine how closely the model emulates or is likely to emulate the real system. This process is known as *validation*. We can put this in a different way by stating that validation is a process of establishing that the model available is a sufficiently correct model of the real system. This does not mean that the model has to be correct down to the last detail. What is required is that the model should be correct in all the areas of operation *that matter*.

This is one of the most difficult areas of simulation study, both in the acquisition of real system data to corroborate the simulation results, and in defining the extent of the data. There are two different aspects from which to view the problem of validation. The first is the case where validation is required for a simulation model of an existing system. The second, but more difficult, problem is validation of a model for a hypothetical or projected system. Let us look at each of these in turn.

Difficulties may be considerable even when there is a real system with which to make comparisons. At least in that case, one may be able to gain access to past operation data for the system. If that is true, one may try to structure a series of simulation experiments to replicate the exact operating conditions that gave rise to the real operation data. This in itself is an extraordinarily difficult process. Imagine being presented with masses of data from the system and from the model. Where do you begin to make comparisons? How do you determine the adequacy of the operating range of the real system from the operation data available? Does the operation data relate to the structure of the model, or to the model parameters, or to both? Questions like these and many more may be asked in relation to 'tuning' a model to make its simulation output responses fit the available data. Control engineers use methods collectively known as *system identification* to try to provide some of the answers.

One of the most valuable outcomes of validation testing of this kind is the fact that we should, in theory at least, be able to improve upon our modelling techniques. Any validation data that can be applied to a model must improve our knowledge of the behaviour of that particular model, and hence improve our modelling techniques in general.

When it comes to trying to validate a model of a projected system that does not yet exist, the task seems impossible. Any validation figures that may be obtained must certainly be incomplete, for by its very nature, most aspects of the real system operation are unknown, since the system does not yet exist. However, we do have recourse to partial validation in the shape of rig tests, pilot plant trials, and other forms of physical testing to fall back on. In Section 1.2.1, physical testing was mentioned more in terms of a form of simulation modelling. As far as simulationists are concerned, one of its main uses is in the partial validation of simulation models. Again it is very difficult to ensure that such partial validation is adequate. Experiments on physical test apparatus for validation purposes have to be carefully designed to give the answers required, to cover the range of operating conditions under consideration, and to enable some judgement as to the accuracy or otherwise of the test results. None of these are easy to obtain, and the outcome of such testing is all too often that the physical test results are more or less useless for the intended purpose. Still such testing may be better than none at all.

If you are interested in reading a very comprehensive account of the problems of validation, you could do well to read the articles by Butterfield and Thomas (1986) and Butterfield (1990). They discuss at length the whole concept of validation from the point of view of both theory and practice, and their thoughts on the subject are of considerable use to the serious practitioner of simulation.

2.5.3 Exploiting the Model

Let us assume optimistically that we have obtained a working simulation, and that we have validated the model and shown that its operation is a true and correct representation of that of the real system. Once again in theory, we should be able to experiment or 'play around' with the simulation model, and be reasonably certain that the real system would respond to the experimentation in a similar way. Experimentation with simulation models is necessary for the reasons given in Section 2.1, so all the checks that are applied in the preceding stages (Section 2.5.2) are a necessary precursor to any experimentation we may wish to do with our simulation model. Inadequacy in any of these stages will vitiate the value of our simulation experiments.

Even if we had achieved perfection in the development and testing of our simulation model, we could not be too sure about experimentation outside the 'boundaries' of our knowledge of the system. In so many cases, the model development we do is aimed at precisely that instance where we wish to try out some new procedure on an existing system and the new operating conditions are outside the range with which we are currently familiar. We should be wary of placing too much reliance on simulation results alone to assist us along this course.

The material in the whole of Section 2.5 shows some of the limitations of simulation as a design and analysis tool. It goes almost without saying, therefore, that as a technique for solving problems, simulation should be applied with a great deal of caution and the results looked at with a very critical eye. If such precautions are not taken, the value of the results obtained will be questionable, and the time and effort spent developing the application will be largely wasted. A very well-known simulation study is outlined below as a classic example of simulation methodology.

2.6 WORLD DYNAMICS – A PREDICTION OF THE FUTURE FOR MANKIND

2.6.1 A Short History

In April 1968, an international group of industrialists and researchers met to discuss the future of the human species in relation to a finite earth habitat. Out of this meeting was formed the Club of Rome. This was a body of private individuals from various backgrounds who shared a common concern, namely the impact of the human population on the planet Earth. In July 1970, J W Forrester of the Massachusetts Institute of Technology (MIT) demonstrated a prototype world model *WORLD2* to the Club of Rome at the start of a programme of work on the Predicament of Mankind. This led to the commencement of a project to develop the model *WORLD3* at the instigation of the Club of Rome. A book (Forrester, 1973) was subsequently published to describe the development of the WORLD2 model.

An early outcome of the Club of Rome project was the publication of a paperback called *Limits to Growth* (Meadows *et al.*, 1972). This was intended to be an easily readable account of the WORLD3 model development, and received considerable world-wide publicity, as did the implications of the model itself for the future of mankind. The project culminated in the publication of a detailed description of the research that led to the WORLD3 model (Meadows *et al.*, 1974). Since then, many other research groups around the world have worked on similar but more complex projects, using the WORLD2 and WORLD3 models as foundations for their efforts.

2.6.2 A Brief Description of the World Models

The basis of each of the world models is a set of mathematical relationships that can best be described as a mixture of differential and algebraic equations together with a number of special functions to quantify empirical data within the models. WORLD2 is the simpler model, having four state (differential equation solution) variables and one algebraic variable that form the heart of the model. These are:

- Human population
- Pollution
- Capital investment
- Natural (i.e. non-renewable) resources
- Food production (as represented by capital-investment-in-agriculture fraction).

WORLD3, having built on the foundations laid by WORLD2, is a more elaborate model which treats the capital sector as being in reality two separate sectors, and expands the food production sector into a much more realistic sub-model. The main variables of WORLD3 are:

- Human population
- Persistent pollution
- Non-renewable resources
- Industrial capital
- Service capital
- Arable land
- Land fertility.

Figure 2.3 shows a simplified block diagram representation of the WORLD3 model, adapted from illustrations in the book by Meadows *et al.* (1974). The seven main model variables are interconnected through a series of other model variables, the cause and effect relationships being indicated by the (+) and (−) signs. For example, increase in available *Health services*, according to the model, is likely to decrease the *Mortality* figures. Increase in *Population* decreases the *Industrial output per capita* since a given *Industrial output* must be shared among more people.

The WORLD3 model was written in the *DYNAMO* simulation language, at that time the preferred simulation medium for system dynamics work at MIT. Initial development of the model was geared towards what became known as the 'Standard Run' – a simulation predicting what the future of the world would be like up to the year 2100, if world policies regarding population growth, exploitation of non-renewable resources, industrialization and pollution were

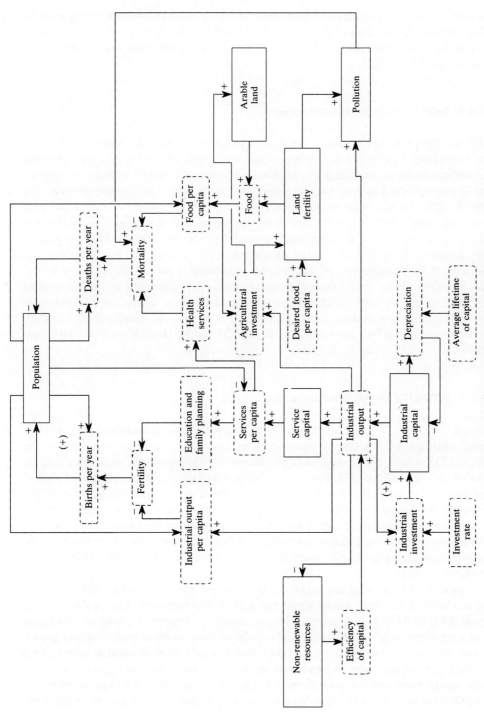

Figure 2.3 Simplified block diagram of WORLD3 model.

allowed to proceed unchanged and unchecked. A full DYNAMO listing is to be found in the Appendix to Chapter 7 of the book by Meadows *et al.* (1974). Figure 7.1 in the same book is a block diagram illustration showing the full complexity of the WORLD3 model.

2.6.3 WORLD3 Model Development

Briefly, the philosophy behind the development of each of the world models was to formulate a plausible set of equations and empirical relationships. The end result was to be a simulation model whose pattern of behaviour could be tailored to fit the historical record averaged over the world, for a set period of time. For WORLD3, the aim was to obtain a reasonable fit to world behaviour patterns over the years 1900 to 1970, and then on the assumption that the model was sufficiently correct, to extrapolate the model timescale beyond 1970 to yield a prediction of future trends for the model variables. Using the ideas developed in Section 2.5, formulation of the model equations and relationships was the *development phase*, fitting the simulation behaviour to observed world history was the *testing phase*, and extrapolation of the simulation timescale beyond 1970 constituted the *exploitation phase*. Full details of the model development work are to be found in the two books on the WORLD3 model (Meadows *et al.*, 1972 and 1974).

2.6.4 WORLD3 Model 'Standard Run'

Figure 2.4 shows a set of curves which represent simulation transient responses or time histories for several of the important main and auxiliary variables in the WORLD3 model. These responses are shown plotted against simulation time, for the years 1900 to 2100 AD. For ease of presentation, the curves are shown plotted on one set of axes, the *X*-axis for time and the *Y*-axis for a scaled value of the variable. Since the variables all have widely disparate values, they would normally each require a separate *Y*-axis scale. In order to keep the plot simple, scaling each variable to a perceived maximum value is a better option, since it is the variable trends that are considered important, rather than their absolute values at any particular point.

In order to interpret the results of the run, we need to consider the responses in the light of what we know about the model development. The period of time from 1900 to 1970 is based on an interpretation of historical evidence, averaged out across the world as a whole. This tells us that world population has grown more or less exponentially, over that period, and that industrial and service output per capita have grown in a similar fashion. This would indicate that industrialization and the provision of services have grown faster than population, and that the world has perhaps been better off for this, as shown by the increasing

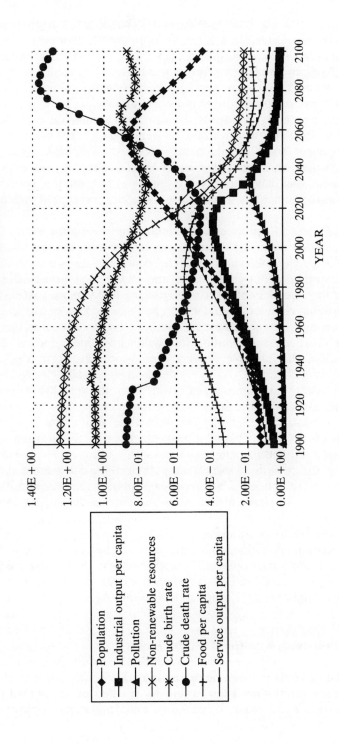

Figure 2.4 WORLD3 model – 'Standard Run'.

food per capita and the falling death rate. The drop in death rate has been perceived, and has resulted in a decrease in birth rate. However, the continuing difference between the two has been responsible for the exponential growth in population. Non-renewable resources, starting from an assumed level in the year 1900, have dwindled somewhat, as resources have been exploited at an increasing rate to meet the needs of a growing industrial base.

If we are comparing trends for the displayed simulation variables over the period 1900 to 1970 AD with world-wide trends, then there would probably not be much argument about their general agreement. We might argue about the degree of absolute agreement that exists between history and simulation, but since the model was developed by matching the two for this period, we must assume that agreement is good enough to constitute a sufficient validation of the model.

When we come to extrapolate the simulation beyond the year 1970, we are sailing in uncharted waters. The simulation results from the 'Standard Run' can be interpreted in the following way. The continuing growth of industrial and service sectors brings about an increasing rate of use of non-renewable resources, as shown by the steepness of the curve reaching a maximum past the year 2000 AD. As remaining resources dwindle even further, the cost of extraction of these resources increases, leaving less capital available for investment in new industries and services. This has a knock-on effect upon investment for agriculture, and would jeopardize the intensive farming upon which the world food supply has become increasingly reliant. As a result, food per capita drops to starvation levels, and the death rate climbs rapidly from about the year 2020 onwards. When the death rate exceeds the birth rate, the population starts to fall back from the peak reached between 2040 and 2060 AD. The death rate is also probably influenced by the drop in industrial growth which affects the provision of services, including education, housing and health care. To a lesser extent, the death rate is influenced by the growth in persistent pollution, resulting from industry and injudicious intensive farming. This reaches a peak between 2020 and 2040, lagging somewhat behind the peaks in industrial and service output per capita, which peak earlier.

The conclusion to be drawn from the simulation is that if world policies towards population, pollution, industrialization and resource extraction remain unchanged, a world catastrophe will occur sometime around the middle of the next century, with death rates unparalleled in history except perhaps by the bubonic plague that ravaged Europe in the Middle Ages.

2.6.5 Experimenting with the WORLD3 Model

In their accounts of their work, Meadows *et al.* (1972 and 1974) tried to answer many of the criticisms they knew would be levelled at the WORLD3 model. Many questions of the 'what if' category were posed and simulation results produced to try to provide answers. We can examine the procedure involved with respect to one important question that has indeed been raised many times. The WORLD3 model assumed a starting level of non-renewable resources in the year 1900 AD. In the light of new discoveries of oil and minerals that are being made, together with more efficient methods of extraction, it could be postulated that the model is too pessimistic regarding the effect of dwindling resources upon the existence of mankind.

To try to give an answer to this criticism, we may pose the question: What would happen if we were to assume that the starting level of resources was double the value originally assumed? We could also look at what we may term the sensitivity of the model behaviour to this starting value by considering a starting level of resources three times the original estimate, to see what effect the changes might have.

Figure 2.5 shows the model responses for the case where the initial level of resources is double that of the 'Standard Run'. The scaled value in the year 1900 is no different, but it now represents twice as great a value in absolute terms. We are now making an assumption about model behaviour that we did not do for the 'Standard Run'. We are, in effect, changing the course of history by choosing a new starting level for non-renewable resources. We then assume that the model behaviour can be relied upon to tell us how different things would have been between 1900 and 1970 if we had had twice as much in the way of non-renewable resources. Then we carry on with the assumption of the 'Standard Run' that we may still extrapolate beyond this point to the end of the 21st century.

If we look at the simulation results shown in Figure 2.5, we can see that the rate of usage of resources appears to have slowed. This is because of the higher starting point. It takes longer to reach the crisis point at which extraction costs become prohibitive, before which time, industrial and service sector growth continue for longer, peaking between 2030 and 2040 at higher levels than before. As a result, pollution peaks between 2060 and 2070, at a level about four times higher than in the 'Standard Run'. It would appear that pollution has a much more significant effect both on food production and directly on death rate, for the peak death rate occurs at the same time as does the nadir of the curve for food per capita. All that would happen, according to the simulation, is that the exponential growth of human population would peak at the same point in time, at a higher value, through being sustained by prolonged industrial and service sector growth. The catastrophe, when it came, would be more intense, probably because of the vastly increased level of pollution and its direct effect upon death rate.

Figure 2.6 shows a repeat of the same simulation, but this time with a three-fold increase in the assumed starting level of non-renewable resources. Once again, the main effect is further prolongation of industrial and service sector growth as a consequence of the abundance of resources. This, in turn, brings about a further increase in the peak pollution and a slight increase in the peak death rate. The population behaviour is quite similar to that in Figure 2.5, peaking at the same value at the same time.

Non-renewable resources initial value = 2.0E12

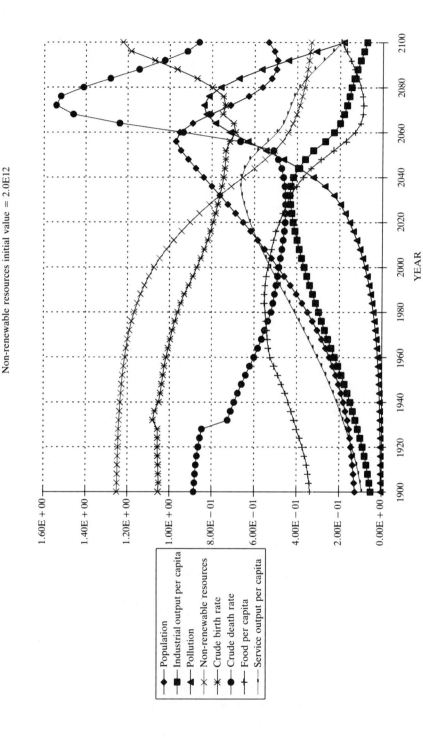

Figure 2.5   WORLD3 model – run with double the initial value of resources.

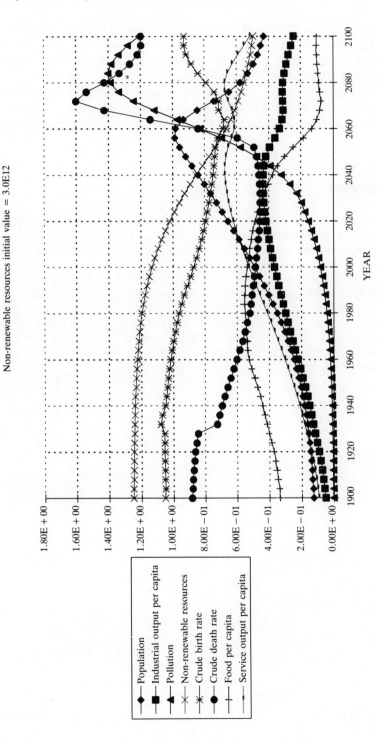

Figure 2.6   WORLD3 model – run with treble the initial value of resources.

Comparison of the results of the three runs would indicate that the main result of increasing the abundance of non-renewable resources is to enable industrial and service sector growth to proceed for longer, with the result that persistent pollution, which is a delayed effect of industrialization, grows to the point where its influence on death rate is likely to exceed the effect of starvation seen in the 'Standard Run'. When assuming that resources are more abundant, there is little to choose between the results. The outcomes are very similar.

2.7 REFERENCES

BUTTERFIELD, M H (1990) 'A method of quantitative validation based on model distortion', in Special issue on model validation, *Trans. Inst.MC*, vol. 12, no. 4, pp. 167–73.

BUTTERFIELD, M H and THOMAS, P J (1986) 'Methods of quantitative validation for dynamic simulation models', *Trans. Inst.MC*, vol. 8, no. 4, pp. 182–219.

FORRESTER, J W (1973) *World dynamics*, 2nd Edn, Cambridge, MA, Wright-Allen Press.

MEADOWS, D H, MEADOWS, D L, RANDERS, J, BEHRENS, W W (1972) *The limits to growth*, London, Earth Island.

MEADOWS, D L, BEHRENS, W W, MEADOWS, D H, NAILL, R F, RANDERS, J, ZAHN, E K O (1974) *Dynamics of growth in a finite world*, Cambridge, MA, Wright-Allen Press.

STRAUSS, J C (Ed.) (1967) 'The SCI Continuous System Simulation Language (CSSL)', *Simulation*, vol. 9, pp. 281–303.

3

Discrete Simulation

3.1 DISCRETE SIMULATION CONCEPTS

If you are unfamiliar with discrete simulation in any form, then the use of the game of MONOPOLY as a simple example of this mode of simulation will have served as an introduction to the subject. It is totally different in concept and in application to the continuous simulation that has been the backbone of the discussion so far. Discrete simulation is very much the province of management scientists and operational research specialists, and is used for addressing the types of problems found in their particular areas of work. Generically, these are all known as *queuing* problems, for they have one thing in common: they are concerned with the delays experienced by objects while waiting to be processed within a system. As a corollary, another goal of this kind of analysis is to try to assess the degree of utilization of expensive resources and facilities that carry out the processing within the system.

Looked at from the point of view of continuous modelling and simulation, the discrete equivalents do not appear to be very relevant. It is easy to see the two sets of techniques as being so different that they could never have anything to do with each other. Nothing could be further from the truth. Discrete modelling has a very important part to play in the continuous scene. As we shall see in Chapter 8, the judicious application of discrete concepts to continuous simulation using digital processors can yield considerable benefit in terms of alleviating some of the difficulties associated with numerical integration over discontinuities. In order to be able to apply discrete concepts in this way, we should have some idea of the methodology involved.

3.1.1 Elements of a Discrete Model

The definitions of a system given in Chapter 1 point to the nature of a discrete model. The objects or components of a discrete system model are usually known as *entities*. These entities are all discrete objects, each being separate from all the others. The entities possess *attributes* which are qualities, characteristics or properties that affect the behaviour of the entities within the model. Entities may be in two kinds of state: either they are *busy*, engaged in some *activity*, or they are *idle*, doing nothing but waiting in a *queue*. A queue in this context does not literally mean a number of entities lined up one behind another. It means a common state in which a number of entities find themselves when they have finished one activity, but before moving on to the next. When an entity moves from an activity to a queue, or vice versa, the state of the system changes. This instant of change is known as an *event*. This is the element of discrete modelling from which its name is derived.

3.1.2 Dynamics of a Discrete Model

The movement of entities through the different points of the model environment constitutes the dynamical behaviour of the model. In order to get this movement to take place, something has to 'galvanize' the system into action. If it were left to sets of different kinds of entities within the model, it is likely that even if there were some dynamical behaviour at the outset, everything would come to a standstill sooner or later, once the stocks of entities for processing were used up. One of the salient features of discrete modelling in the context of most applications is the fact that sustained dynamical behaviour within the model depends on the movement of entities of different types. These will move into the model environment, possibly at different points, and depart from the model, again possibly at different departure points. These *arrivals* within the model environment are what keep things going. If we borrow some of the terminology of continuous simulation, we may call these arrivals within the system the *forcing functions* that make the system model behave dynamically. There are different approaches to the implementation of these concepts as discrete models, but there are two features that are common to all, in terms of simulation time advance and the accompanying change of state of the system model.

The *events* are the instants at which the state of the model changes through the displacement of entities between points throughout the model environment. The state of the model remains constant at all other points in time. If, therefore, we can keep track of all the events in the system, we may set up a mechanism for simulation *time advance*. This is brought about by ordering all known events into a chronological order of occurrence, and letting simulation time advance from one event to the next in the ordered sequence. Since the state of the model is constant between events, nothing is lost by skipping from one event to the next with no consideration of the time in between the two. We can now consider the different modelling approaches that implement this time advance mechanism.

3.2 THE DIFFERENT MODELLING APPROACHES IN DISCRETE SIMULATION

Three different approaches are in vogue at the present time for the modelling of discrete systems. The first two are more or less universally popular, but particularly so in North America, while the third approach has found more favour in Europe. From the short descriptions of each given below, you are free to choose which of the three modelling philosophies suits you best. The differences between each method lie in the way in which the model is activated and kept going in each case. A very simple modelling example will illustrate the differences. A queue of people waiting to use a single public telephone constitutes a system that may be modelled by discrete simulation methods. The justification for carrying out a modelling and simulation exercise on such a system would be the need to obtain answers to various questions concerned with the operation of the facility. Typical questions would be:

- Do people have to wait a long time on average?
- Do long queues predominate?
- Is there a need for another telephone booth?

If one had access to data concerning the arrival rates of would-be callers, and information concerning the average duration of calls, then implementation of a simulation incorporating these statistics would be a relatively simple task, capable of giving answers to all the questions above. Information of this nature is available from some sources, notably the telephone companies, and does get used to provide the answers above.

3.2.1 Entity-Oriented (Process Interaction) Models

As the name suggests, this mode of operation emphasizes the entities entering the model environment. It is concerned mainly with following their progress throughout the model, recording all changes, delays, waiting and working times incurred by these entities, until they finally leave the bounds of the model. Statistics are gathered, amongst other things, for the average waiting times in all the queues, for the numbers of entities processed by the model, and for the utilization of facilities used to process the migratory entities (often called *transactions* in the jargon of this mode of modelling). Figure 3.1 illustrates this modelling technique applied to the telephone queue. The transactions of the system are the people who arrive, join the end of a queue, move up towards the telephone booth, make calls and leave the model. The facility is the telephone booth which is always within the model. Each person arriving automatically becomes labelled with an arrival time. The length of time spent waiting in the queue is recorded and added to the overall statistics. The time spent making a call is also recorded and added to the statistics from which the average utilization of the booth is calculated, as well as the average length of a telephone call.

By way of a specific external example of the application of this technique, one form of demographic modelling involves the use of the so-called 'cohort' method for modelling change of population within human society. The population is divided into groups more properly called *generations* but sometimes known as *cohorts*, which enter the population model at time zero (i.e. at birth). Each cohort

Figure 3.1   Public telephone queue – entity-oriented model.

is allowed to age in the normal way, and records are kept of depletion each year
within the cohort, brought about by mortality rates being applied at each age. The
progress and size of the cohort are recorded until it disappears altogether when
the last members die. That is the essence of the model, in that the entity entering
the environment – the cohort – is the important feature. Its progress is followed
until it disappears (i.e. leaves the model bounds). On its way through the model,
various interactions have to be brought into play to generate the population
members for new cohorts, to keep the birth/death process going. Figure 3.2 shows
a graphical form known as a 'population pyramid', often used by demographers to
display the relative cohort sizes in male and female populations. The figures on
this graph are for a population with relatively constant mortality rates, with higher
than average rates in infancy, early adulthood (caused by armed conflict for
males, and childbirth for females), and old age.

Entity-based modelling is relatively easy to learn and apply as a concept, but
it is also considered to be rather limited for the modelling of complex systems,
since it does not have sufficient flexibility.

3.2.2 Event-Based Modelling

The mechanism of this type of modelling causes model behaviour to depend upon
the events that occur during the dynamic execution of the simulation. Entities
'arrive' within the model environment, and move through to their discharge
points. The model events are the means whereby the movement of entities is
mechanized from one point to another. When an event time is reached, two
actions have to be fulfilled. Firstly, the event itself is processed, which means that

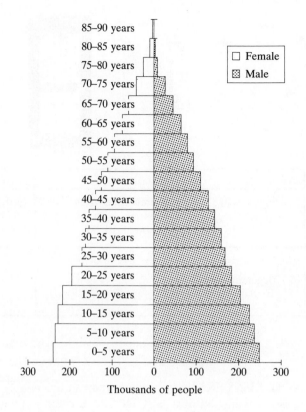

Figure 3.2   Population pyramid – example of cohort modelling.

whatever actions are scheduled to take place at that instant of time are accomplished. Secondly, since event times are the only points at which changes of state are allowed, the scheduling of further events must then occur. In this way, an interactive sequence of events causes the model to behave dynamically.

Figure 3.3 shows the telephone queue modelled from an event-processing point of view. The events that 'drive' the system are the ARRIVALs of would-be callers to join the queue. From knowledge of the average rate of arrival, a random sequence of ARRIVALs is generated as a series of events. Each ARRIVAL schedules the next in sequence, the interarrival interval being sampled from a statistical distribution. The processing of the event is simply the generation of the caller to join the queue.

The further logic of the model depends upon the circumstances. If there is no queue, the ARRIVAL will also schedule the BEGINning of a CALL, which takes place immediately upon arrival. The BEGINning of the CALL generates the LEAVE event that signifies the end of the call. A sampled call duration will yield

EVENT	ARRIVAL	BEGIN CALL	LEAVE
PROCESS THE EVENT	Generate caller to join queue	Extract caller from queue to begin call. Set booth BUSY	Extract caller from booth. Set booth IDLE
SCHEDULE ANY OTHER EVENTS	(1) Next arrival (2) Beginning of call (if no queue exists)	Leave at end of call	Beginning of next call (if queue exists)

Figure 3.3   Public telephone queue – event-based model.

that figure. Once the LEAVE event is reached, the call terminates and the caller departs. If there is a queue, the next BEGIN CALL event is scheduled, and so the process is maintained all the time that people arrive to make calls.

An everyday example for which event-based simulation is very suitable is an inventory or stock control system. Maintaining stocks of even a single product line costs money. Stock represents capital tied up in the value of the goods. The greater the stock, the larger must be the warehouse space to carry it, and the greater will be the carrying charges. The ideal situation will be that in which minimum stocks are carried by operating on a *just-in-time* basis. This means that fresh deliveries of the stock items are made to the warehouse just before stock runs out, and that stocks are maintained at a permanently low level.

Against this must be offset the loss of sales and potential goodwill if stock does run out, and the need for frequent stock-taking if permanently low levels of stock are to be feasible. Both of these also cost money, so the design of a stock control system involves balancing one set of charges against the other in such a way as to minimize the overall costs of running the system.

In order to construct an event-based model for such a system, we have to define all the possible types of event that we consider could take place in the operation of a stock control model. When we have done that, the next stage is to set out clearly just what actions will be fulfilled, according to type, when each

Table 3.1 Event processing interaction for stock control system

Event	*Process the event*	*Schedule the next event*
SALE of stock item	Stock item leaves the warehouse (and the model). Ignore if stock has run out	Next SALE event
STOCK-TAKING	Take stock of the goods in the warehouse. Place order if stock is below the re-order point level	Next STOCK-TAKING DELIVERY of goods, if an order was placed
DELIVERY of goods	Receive delivery of goods. Add to stock	No scheduling of other events

event is processed. Table 3.1 shows what events are considered necessary for our model of a single product line. Once these events have been identified, the requisite interactions between them can be mechanized to construct the simulation model. An event table (which is itself a queue of events set in chronological order) will be at the heart of this model, and it will behave dynamically, having events added to it in the correct places as they are scheduled, and removed once they have been processed.

3.2.3 Activity Cycles

The last of the three modelling approaches to discrete simulation that we should consider is the activity-based or activity-scanning mode of operation. As its name suggests, the method is centred around the activity, which we may define as a busy state of entities, bounded by two events. These are a *start-activity* event and a *stop-activity* event. Since an activity is a busy state of entities, this implies that it cannot take place unless all the entities required are available. Furthermore, it also implies that the entities involved in an activity cannot be released until that activity has ceased. At each event time, the activities are scanned to ascertain if any are in a position to start. We may infer from the occasion of the start-activity event that a change of state takes place in the associated entities, from idle to busy. The same may be said about the stop-activity event, where a change takes place from busy to idle. We may also infer, therefore, that the activity must be preceded and succeeded by at least one queue in each case.

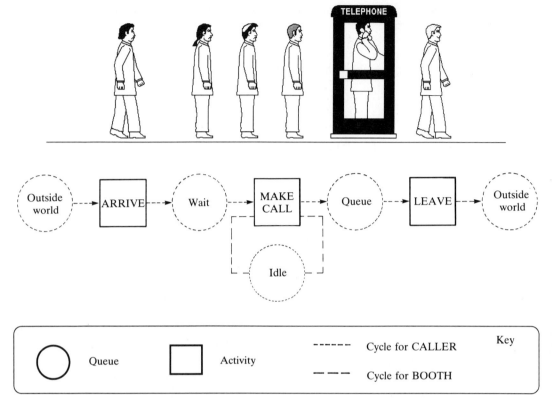

Figure 3.4 Public telephone queue – activity life cycles.

The above reasoning gives rise to the concept of the activity cycle, which is nothing more than an alternating sequence of queues and activities, starting and ending with a queue. If the same queue is used for the start and end of the cycle, a closed cycle results.

The notion of entity 'classes' enters into the activity cycle model of a system. We need to define a separate cycle for each class of entity that exists, so the starting point for this type of model is consideration as to just what kinds of entity it should contain. Once we have done that we may draw up the corresponding activity cycles. Cycles will be connected through activities that require entities from more than one class. In this way large and complicated models may be built up.

Figure 3.4 shows the queuing system for a public telephone, modelled by means of activity cycles. There are two cycles: one for the class of entity that we may refer to as CALLER, and the other for the BOOTH. These are the only entity classes present in the system, so the two cycles are sufficient for the model. CALLERs enter the model environment by means of the ARRIVE activity. The queue that defines the boundary is labelled 'Outside world', as is the queue on the other side where the CALLERs LEAVE the model. A closed cycle could therefore be used for the entity CALLER, if so desired.

The alternating sequence of queues and activities within a cycle serves to give this modelling approach a useful 'syntax' or set of rules to apply when constructing models. Using the telephone model as an example, some of these may be summarized as follows:

- The alternating sequence of queues and activities is strict.
- The activity MAKE CALL is a *conditional activity* that can only commence if both kinds of entity necessary for its operation are available.
- The activity LEAVE is a *bound activity* that will start as soon as MAKE CALL has finished.
- The queue in between these two activities is a dummy queue, there only to satisfy the syntax.
- The ARRIVE activity – strictly an event – is treated as an activity of zero duration.
- The cycle for the BOOTH is a closed cycle, connecting with that for the CALLERs through the common activity MAKE CALL.

Activity cycle modelling has found considerable application to the simulation of production-line processes. The type of system where there are one or more production lines, with entities passing along them from one manufacturing stage to the next, is ideally suited to this method of modelling. The requirement of entity availability for activities to commence enables ready consideration of the effects of machine breakdown and operator absence on the production process.

3.3 DISCRETE SIMULATION SOFTWARE – THE BUILDING BLOCKS

Although we are not going beyond the principles of discrete simulation, it is expedient for us to take a look at some of the intrinsic features of discrete simulation packages in a general way, looking ahead so that we will know what aspects will be of use in our promised application to continuous simulation. Some features will be readily identifiable as being of use, others will seem to be irrelevant.

3.3.1 Model Definition and Set-up

From the examples given in Section 3.2, we have seen how the three modelling approaches have their different starting points. The entity-based (process interaction) and activity modelling methods both concentrate upon the entities of the system as the starting points. This is to be expected in the case of process interaction, but the need for a cycle for each entity class in activity modelling leads to the same starting point. Event-based modelling is unique in that we would start by trying to define all possible events that might come to pass during the operation of our system model.

Whichever starting point we may use, after due consideration of the basic elements of the model, we should have a clear idea of what entities, events, activities and queues should be included in the model. We should only include model elements if we can visualize that they appear to be indispensable for the proper operation of the model. For instance, in the telephone simulation, the telephone instrument itself is not considered to be an entity separate from the BOOTH. This is because we have assumed that the availability of the instrument is inseparable from that of the BOOTH, so that they may be treated as one entity.

In Section 3.1.1, the term *attribute* was defined as a property of an entity. In modelling terms, entities should possess attributes only if required for the proper functioning of the model. The CALLERs in the telephone queue might be given the attribute MARRIED or SINGLE if by doing so one could show that marital status had a bearing on the duration of the telephone calls that would subsequently be made!

The method of setting up a model on a computer is very much dependent on the simulation package being used. Some packages require the input of a section of language code to describe the model features. Together with data, this code is processed by a language translator to provide a run-time module that will execute in a standard way. Other packages provide an interactive input facility, whereby model implementation is taken one step at a time under the control of suitable software. This checks the model for errors and inconsistencies before carrying out the simulation, thereby easing considerably the pain of 'getting it right'.

3.3.2 Statistical Distributions

In Section 1.3, we established the association of random variability with discrete simulation. We should now consider how that variability can be implemented in a computer-based model.

We have seen how, in the telephone model, there can be two random processes at work. The first one is the arrival of the callers to join the queue. One has only to watch the real situation in action to realize that people do not join a queue at regular intervals. The arrivals are haphazard, and there appears to be no easily defined way of modelling the arrival process. The second random variation takes place in the duration of the telephone calls. Again there seems to be no way of determining in advance just how long a particular call should be. Observation of a real situation would bear this out.

We are fortunate, in discrete simulation, that we have one factor going for us. It is seldom the case that we need to be able to point to a caller in the telephone queue and state just what the duration would be for the corresponding telephone call. The information we seek from a discrete simulation is in the form of *average patterns of behaviour* for the various elements of a model. This means that individual patterns of behaviour are unimportant, especially if the number of entities under consideration is very large. It also means that we are free to generate random variability by sampling from the appropriate statistical distribution that best fits the random characteristics of the system we are trying to model. By so doing, we replicate the average behaviour patterns, without having to pay any attention to individual characteristics.

Without going much into the theory of probability and statistics, it is useful for us to examine in a qualitative way some of the more important and useful statistical distributions that are used to replicate random variability in real world situations.

3.3.2.1 General probability concepts

Figure 3.5 shows a diagram of what is known as a *probability density function (PDF)*. What this states may be put simply as follows: the probability that a value x will occur is given by $P(x)$ at any point, expressed as a fraction. The probability that x will have a value between a and b is given by the area under the curve of $P(x)$ that lies between these two points. This will also be expressed as a fractional value. The probability that x will have a value between $-\infty$ and $+\infty$ is the total area under the curve, and since it encompasses all possible values for x, the probability will have a value of 1.0, in our fractional terms. The PDF is not very useful in terms of its application to discrete simulation. What we will find more useful, as we shall see later when we apply it to sampling from distributions, is the

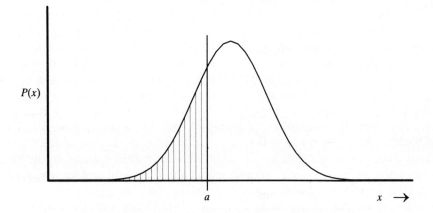

Figure 3.5   General probability density function.

cumulative density function (CDF). For a value a within the range of all possible values of x, the CDF represents the total probability that x will lie somewhere within the range $-\infty$ to a. In mathematical terms,

$$C(x)|_{x=a} = \int_{-\infty}^{a} P(x)\,dx \tag{3.1}$$

This integral is represented graphically by the shaded portion of the area under the curve for $P(x)$ shown in Figure 3.5.

3.3.2.1 General probability concepts

This distribution may arise, for example, in the study of rounding errors when measurements are recorded to a certain degree of precision. If we were to carry out a series of measurements of weight, in which we recorded the results to the nearest gram, we might be able to assume, for one measurement, that the difference between the actual weight and that which we had recorded was some number between $-0.5\,g$ and $+0.5\,g$, and that the error was uniformly distributed throughout this interval.

For a uniform distribution, the probability of a value occurring between the limits a and b is given by the probability density function (PDF):

$$P(x) = \frac{1}{b-a} \qquad (a<x<b)$$
$$= 0 \qquad \text{(otherwise)} \tag{3.2}$$

The cumulative distribution function (CDF) is

$$C(x) = \frac{x-a}{b-a} \qquad (0 \leqslant C(x) \leqslant 1.0) \tag{3.3}$$

These functions have the form shown in Figure 3.6(a) and (b) respectively.

A much more important use for the uniform distribution is the generation of random numbers. In simulation, as we shall see later, we should be able to generate random numbers as a series of samples from a uniform distribution, where the range of numbers generated lies between 0 and 1.0. If the numbers are truly random, they will be equally likely to occur anywhere within the range, and the PDF will be shaped as in Figure 3.6(a). In this way, we may regard the generation of random numbers as a special case of sampling from a uniform distribution where the range is 0 to 1.0. If we were to require samples from a uniform distribution with a different range, then the samples drawn would be the random numbers we generate, suitably scaled and offset in the following way.

If R is a random number generated with a value between 0 and 1.0, then $(b-a)R$ will be the sample from a uniform distribution with the range $(b-a)$ where the bottom end of the range is zero. If the lower end of the range is a, as

(a)

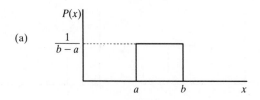

(b)

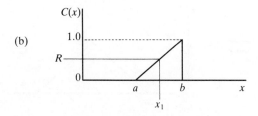

Figure 3.6 (a) PDF for a uniform (rectangular) distribution; (b) CDF for a uniform (rectangular) distribution.

we would expect from the range $(b - a)$, then our sample from the corresponding uniform distribution will be given as

$$x_1 = a + (b - a)R \tag{3.4}$$

On the CDF diagram in Figure 3.6(b), we may interpret this geometrically as follows. Let the Y-variable $C(x)$ represent the range of all random numbers between 0 and 1.0. For any value R lying within this range, we may draw a line parallel with the X-axis to intersect the Y-axis at R. We now produce this line so that it intersects the CDF for the uniform distribution, and drop a perpendicular from that point of intersection. The corresponding value read off the X-axis is the required sample x_1 from the uniform distribution. In the jargon of statistics, this sample from a statistical distribution is known as a *variate*.

With any statistical distribution, two parameters that we normally need to specify are the *mean*, *average value* or *expected value* and the spread about the mean given by the *standard deviation* or, more properly, by the *variance* which is the square of the standard deviation. For the uniform distribution, these are given by the formulae

$$\mu = \frac{b + a}{2} \qquad \text{(mean or expected value)} \tag{3.5}$$

$$\sigma^2 = \frac{(b - a)^2}{12} \qquad \text{(variance)} \tag{3.6}$$

On a lighter note, if we return for one moment to board games, then we may

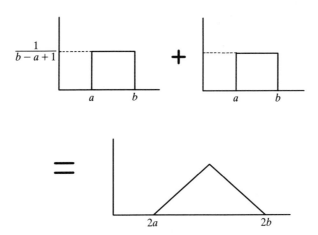

Figure 3.7   Formation of a triangular distribution.

consider the 'random-number generator' that we use in this context (i.e. a single die) as a discrete form of the uniform distribution, where

$$P(x)_i = \frac{1}{(b-a)+1} \quad (i \in a, a+1, a+2, \ldots, b) \tag{3.7}$$

$$C(x)_a = 0 \qquad C(x)_b = 1.0$$

For example, if we throw one die to obtain a number between 1 and 6, then if it is a 'fair' die,

$$P(x) = 1/6 \quad \text{for all numbers equally}$$

3.3.2.3 Triangular distribution

We may now use the concept of a uniform distribution to obtain another form of distribution. If x and y are uniform variates (this means that they are each samples from uniform distributions), then if

$$z = x + y \tag{3.8}$$

z will be a sample from a triangular distribution. The principle behind this derivation is illustrated in Figure 3.7. In discrete form, the game of MONOPOLY gives us a good example of sampling from a triangular distribution. If we throw two dice together, the sum total of the scores on both dice is a sample from a triangular distribution whose range is 2 to 12. The apex of the triangle coincides with a value of 7 (which is the most likely outcome of the throw).

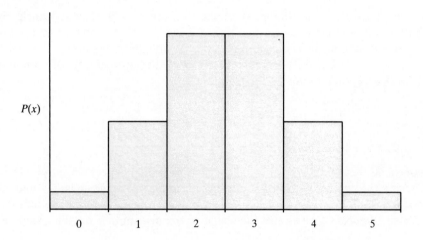

Figure 3.8   Histogram of a binomial distribution.

3.3.2.4 Binomial distribution

This is a distribution that is useful in the field of quality control, where the sampled elements have two kinds of attribute (e.g. yes/no responses, acceptable/ defective items, etc.). The binomial distribution is a discrete distribution that gives the probability that an event or a 'success' occurs x times out of n trials where the overall probability of success has a value p (and hence an overall probability of failure given by q where $q = 1 - p$). The PDF for the binomial distribution is given by

$$P(x) = \binom{n}{x} p^x q^{n-x} \quad (x = 0, 1, 2, \ldots, n) \tag{3.9}$$

We may state the normal statistical parameters as

$$\text{Mean or expected value} = np \tag{3.10}$$

$$\text{Variance} = npq \tag{3.11}$$

Let us consider the example of an acceptance/rejection situation where the probability of acceptance $p = 1/2$ and the probability of failure $q = 1/2 = 1 - p$. Using the general formula for the binomial distribution given by equation (3.9), we may state that the probability of obtaining $0, 1, 2, \ldots, n$ acceptable items out of a sample of n items in total is given respectively by

$$q^n, \binom{n}{n-1} q^{n-1} p, \binom{n}{n-2} q^{n-2} p^2, \ldots, p^n \tag{3.12}$$

These are terms of the binomial expansion for $(q + p)^n$ from which the distribution gets its name. It may be easier if we substitute some numbers into these formulae so that the relative magnitudes become more apparent. If $p = 1/2$, $q = 1 - p = 1/2$ and n is a sample of 5, then the probability of obtaining $0, 1, 2, 3, 4, 5$ acceptable items in the sample of 5 is

$$\left(\frac{1}{2}\right)^5, 5\left(\frac{1}{2}\right)^5, 10\left(\frac{1}{2}\right)^5, 10\left(\frac{1}{2}\right)^5, 5\left(\frac{1}{2}\right)^5, \left(\frac{1}{2}\right)^5$$

$$= 1/32,\ 5/32,\ 10/32,\ 10/32,\ 5/32,\ 1/32$$

The sum of all these terms is 1.0, which is what we would expect since these terms characterize the whole distribution for this example. The mean is 2.5 and the variance is 1.25. Since this is a discrete distribution, it is more proper to plot it as a *histogram*. Figure 3.8 illustrates the shape of the distribution for this example.

3.3.2.5 Poisson distribution

This distribution is one of the most important in discrete simulation, not so much in terms of direct application, but more by virtue of the application of its so-called 'inverse' – the negative-exponential distribution. More about that later. In the meantime, we may consider what kinds of phenomena are Poisson distributed, just what this distribution is, and how we may derive it in a simple way.

If you have studied physics, and carried out an experiment to determine the frequency of detection of particles in background radiation, then you may have discovered (or been informed) that the count was Poisson distributed. It is interesting to conjecture just why this should be the case. One of the characteristics of radiation is the fact that the arrival of an individual particle is something that has an extremely small probability of occurrence. Also, over the course of time during the experiment, large numbers of detections would have been registered. The conclusion that we may draw from this phenomenon is that, in such an experiment, we would register large numbers of arrivals of particles, each of which individually has a very small probability of happening at all within the time frame of the experiment. There seems to be some kind of connection with the binomial distribution, in that, in the experiment, we would register the arrival of a particle as a success. Let us start with the binomial distribution and see where that leads us in obtaining the Poisson distribution.

Let us consider the flow of vehicular traffic on a road, past one of those observation points much used by road traffic engineers. Let us assume that this flow has an average value of only 1 vehicle per minute, with vehicles arriving at the point at random and independently of each other. Over short intervals of time such as successive minutes, we would not expect exactly equal numbers of vehicles to arrive, least of all 1/60th of a vehicle in each successive second. It is possible, however, that one or more vehicles may arrive in a particular second.

If we let the probability of 1 vehicle arriving per second be 1/60, then a period of 1 minute corresponds to 60 trials of an event with a probability of 1/60, or to 120 trials of an event with probability 1/120. What we have here, therefore, is a situation where the number of trials times the probability of an arrival during each trial period is constant.

We could use the binomial distribution to calculate the probabilities of $0, 1, 2, \ldots$ vehicles per minute, by application of the formulae in equation (3.12) to give

$$q^n, \binom{n}{n-1} q^{n-1} p, \binom{n}{n-2} q^{n-2} p^2, \ldots$$

Now we may write

$$\ln(q^n) = n \cdot \ln(q)$$

$$= n \cdot \ln(1 - p)$$

$$= -np \quad \text{(if } p \text{ is very small)} \tag{3.13}$$

This means that if np is constant, as $n \to \infty$, $p \to 0$, so that

$$q^n \to e^{-np} = e^{-\mu} \tag{3.14}$$

where μ is the mean or expected value. The binomial distribution tends in the limit towards the Poisson distribution. This would reflect the kinds of conditions present in the traffic flow situation in which a given average number of incidents (arrivals at the observation point) is the result of a large number of opportunities for occurrence of an arrival. This large number results from the subdivision of an observation period into very small intervals of time. This, in turn, means that the probability of an arrival taking place within one of these minute time periods is correspondingly small. This situation, as you will remember, is exactly that which characterized the radiation particle counting experiment.

Since $p \to 0$, $q \to 1$

$$\therefore \text{Variance} = npq = np = \mu \tag{3.15}$$

The Poisson distribution therefore tells us that the probability of $0, 1, 2, 3, \ldots, n$ successes is

$$e^{-\mu}, \mu e^{-\mu}, \frac{\mu^2}{2!} e^{-\mu}, \frac{\mu^3}{3!} e^{-\mu}, \ldots, \frac{\mu^n}{n!} e^{-\mu} \tag{3.16}$$

where μ = mean = variance. This property of identical mean and variance is the most important feature of the Poisson distribution.

In order to move towards being able to use this result in discrete simulation, we need to express the Poisson distribution in a different way, in terms of a parameter we need to use in simulation, namely the *arrival rate*. If we call this arrival rate λ, then we may express the corresponding *interarrival interval* as $1/\lambda$.

If we count a single arrival as one of the successes mentioned above for the Poisson distribution, then the mean number of arrivals in a time interval $t = \lambda t = \mu$ (which is the mean of the Poisson distribution). We may therefore re-write the last term in equation (3.16) to give us the probability that x arrivals occur in a time interval t as the equation

$$P(x,t) = \left[\frac{(\lambda t)^x}{x!} \right] e^{-\lambda t} \tag{3.17}$$

This relationship holds if we can satisfy the following conditions:

- Events occur completely independently of each other. This means that, within a time interval, the number of arrivals that take place is not influenced in any way by previous arrivals in previous time intervals.
- The average rate of arrival λ remains constant. This is the concept of a *stationary* distribution.

Sampling from a Poisson distribution within a simulation is not the easiest thing to do. There is, however, a way round the problem if we presuppose the following to be true. If in some way we are able to sample a distribution to give us the interarrival interval as a variate, then a number of such intervals within a time period t as used in equation (3.17) will constitute a sample x from a Poisson distribution whose mean or expected value is λ. At the beginning of Section 3.3.2.5, mention was made of the negative-exponential distribution being the 'inverse' of the Poisson distribution. If the two conditions above for the Poisson distribution hold true, then it may be shown that the PDF of the time interval t between the occurrence of successive arrivals is given by

$$P(t) = \lambda e^{-\lambda t} \tag{3.18}$$

whereas in the corresponding Poisson distribution given by equation (3.17), we have an expression for the PDF of the likelihood of x arrivals occurring within the time interval t. In this way, we have one function as the inverse of the other. As we shall see in the next section, the negative-exponential distribution is by far the easiest to implement within a computer simulation, and therefore Poisson variates should be generated as described above from a series of negative-exponential variates.

3.3.2.6 Negative-exponential distribution

If the rate of occurrence of arrivals is Poisson distributed, then it can be shown that the PDF of the time interval t between successive arrivals is given by

$$P(t) = \lambda e^{-\lambda t} \tag{3.19}$$

This is simply equation (3.18) re-written with the conventional symbol for PDF, and is known as the **negative-exponential** (or sometimes just exponential)

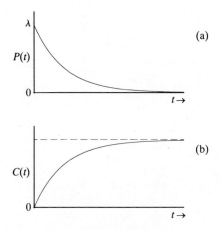

Figure 3.9   (a) PDF for a negative-exponential distribution; (b) CDF for a negative-exponential distribution.

distribution. The corresponding CDF is the cumulative total probability that all such generated time intervals will lie between times 0 and t, and is given by

$$C(t) = \int_0^t \lambda e^{-\lambda t} dt = [-e^{-\lambda t}]_0^t = 1 - e^{-\lambda t} \qquad (3.20)$$

It is interesting to try to visualize just what this means in practice. If we look at Figure 3.9(a), we can see that the PDF for the negative-exponential distribution is a decaying exponential which intercepts the Y-axis at a value of λ, the mean arrival rate. What the curve represents is the likelihood that an interval of a given size t will fit into all the intervals that are generated by sampling. An interval of zero duration will fit into all intervals generated; therefore the PDF for this is the same as the mean arrival rate, λ. When we increase the size of our 'test' interval beyond zero to any value t, it no longer fits into all the generated time intervals, and the longer the test interval, the smaller its chance of fitting into one of the samples. A zero time interval will have the highest probability (λ) of fitting in between successive arrivals. This probability decreases exponentially as the size of the time interval increases.

Figure 3.9(b) shows the CDF for the negative-exponential distribution. Continuing our analogy of a 'test' interval, $C(t)$ represents the cumulative total probability that all the intervals generated will fit into our test interval of size t. As the size of the interval increases, $C(x)$ approaches a value of 1.0 asymptotically.

We may use the same method of drawing a sample from a negative-exponential distribution as we used for uniform variates. If we sample from a uniform distribution with a range between 0 and 1 (i.e. use a random-number generator to do this for us), then we have a value that is equally likely to occur within the range 0 to 1 and we may represent it by a value R on the $C(t)$ axis of

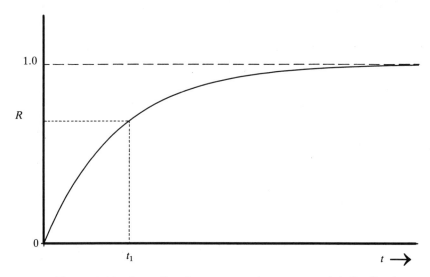

Figure 3.10   Sampling from a negative-exponential distribution.

Figure 3.10. The corresponding value of t produced by the construction shown in Figure 3.10 is a time interval t_1 that is a sample from the negative-exponential distribution. Since R can occur anywhere on the $C(t)$ axis with equal likelihood, the shape of the CDF ensures that smaller intervals are more likely to be generated than larger ones. Fifty per cent of the samples will occur in the lower end of the range, while the remaining fifty per cent will span a wide range of possible time intervals.

In computational terms, it is very easy to draw samples from a negative-exponential distribution. If we generate a random number R where $(0 \leqslant R \leqslant 1.0)$ then

$$C(t) = R = 1 - e^{-\lambda t} \tag{3.21}$$

$$\therefore e^{-\lambda t} = 1 - R \tag{3.22}$$

$$\therefore \lambda t = -\ln(1 - R) \tag{3.23}$$

Since R is uniformly distributed in the interval $(0 \leqslant R \leqslant 1.0)$, $(1 - R)$ is also uniform in the same interval. We will therefore obtain the same effect by using the relationship

$$\lambda t = -\ln(R) \tag{3.24}$$

from which we obtain the working relationship

$$t = -\frac{1}{\lambda} \ln(R) \tag{3.25}$$

where t is an interval of time sampled from a negative-exponential distribution.

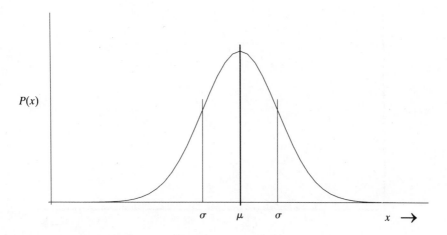

Figure 3.11   PDF of a normal or Gaussian distribution.

3.3.2.7 Normal or Gaussian distribution

This is a continuous symmetrical distribution widely used in computer modelling and simulation. We may use it to describe the random variation in most measurement phenomena, for example:

- Angular or linear measurement errors (not the same as rounding errors which are uniformly distributed)
- Variations in group examination marks
- Variability in human attributes such as height or body weight for given groups.

The PDF for the normal distribution is given by the rather complicated-looking formula

$$P(x) = \frac{1}{\sigma\sqrt{2\pi}}e^{-(x-\mu)^2/2\sigma^2} \quad (-\infty < x < +\infty) \tag{3.26}$$

where

μ = mean or expected value
σ = standard deviation
σ^2 = variance

Figure 3.11 shows the general form of the normal distribution, with the mean value also being the axis of symmetry of the curve. The standard deviation lines are shown to indicate the effect the value of σ will have on the shape of the curve. Large values of σ will cause the curve to be spread out somewhat, while small values will produce a narrow bell shape.

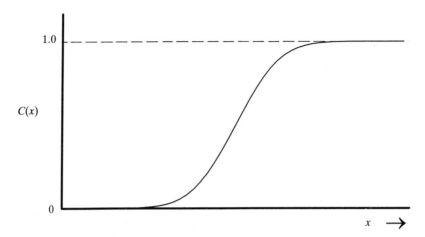

Figure 3.12   CDF for a normal distribution.

We may simplify the PDF formula to some extent by giving special values to the mean and standard deviation. For $\mu = 0$ and $\sigma = 1$, we may write down a formula known as the *standard normal distribution* given by

$$P(z) = \frac{1}{\sqrt{2\pi}} e^{-z^2/2} \quad (-\infty < z < +\infty) \tag{3.27}$$

We may convert any normal distribution into standard form by making the substitution

$$z = \frac{x - \mu}{\sigma} \tag{3.28}$$

The CDF for the normal distribution is given by writing

$$C(x) = \int_{-\infty}^{x} P(x)\,dx = \frac{1}{\sigma\sqrt{2\pi}} \int_{-\infty}^{x} e^{-(x-\mu)^2/2\sigma^2}\,dx \tag{3.29}$$

This has the shape shown in Figure 3.12.

In computational terms, it is no easy matter to try to generate the normal variate x (or z) by inverse transformation using the CDF curve. Instead of this complicated procedure, we may use a simpler method based on gathering together a group of uniform variates (samples from a uniform distribution). It can be shown that if such a series of uniform variates R_k is generated, a standard normal variate z may be derived to a close degree of approximation, using the formula

$$z = \frac{(\sum_{k=1}^{m} R_k) - m/2}{\sqrt{m/12}} \tag{3.30}$$

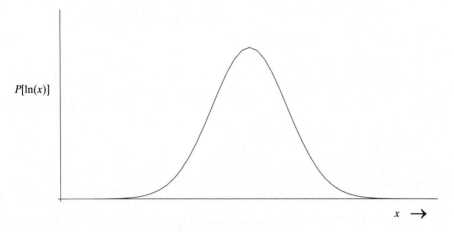

Figure 3.13 $P(\ln(x))$ as a normal distribution.

If $m = 12$, then

$$z = \sum_{k=1}^{12} R_k - 6 \tag{3.31}$$

If we need to convert this sample to one from a non-standard normal distribution, we may make the substitution

$$x = \mu + \sigma.z \tag{3.32}$$

3.3.2.8 Log-normal distribution

The log-normal distribution is simply one in which $P[\ln(x)]$ is normally distributed. This is shown in Figure 3.13 and has a shape no different from the normal distribution considered in the previous section. When, however, we exponentiate this function (take the anti-logarithm) we obtain the true log-normal distribution which has the skewed shape shown in Figure 3.14. By virtue of the properties of logarithms, this distribution can only exist for random variations where the measured values are all positive. You may be wondering what use such a shape could have in describing random variability. There are indeed randomness situations in everyday life which fit this shape of curve quite well. Granular materials that have reached a certain degree of fineness after a grinding operation will in fact contain particles of differing sizes, whose relative abundance may be log-normally distributed. A similar shape too may well describe the distribution of income over the population as a whole. If we consider x as representing income per capita, the shape of the curve for low incomes suggests that few people have really small incomes, perhaps through the medium of a state-run welfare agency.

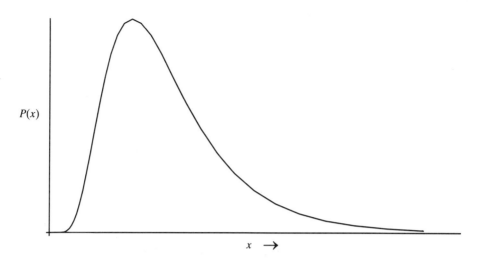

Figure 3.14   *P(x)* as a log-normal distribution.

The steep rise over a relatively small increase in income may suggest rapid inclusion of much of the population into the group as income increases. The skewed peak suggests that the vast majority of the population receive modest incomes. The long tail to the curve may suggest relatively few people enjoy really high incomes, the range of which is far greater than the income average for the population as a whole.

An inverted form of the log-normal distribution is shown in Figure 3.15. To some, this is the familiar 'bathtub' curve that may be used to represent the distribution of mortality with age in populations. In a typical population, a relatively high mortality rate in early life gives way to low mortality during youth and early adulthood. However, mortality gradually increases with advancing age as shown by the tail of the curve. The same notion applies to failure rates of equipment. If we consider the purchase of a new car, it frequently happens that there are many teething troubles that become apparent soon after purchase. Once these initial problems have been ironed out, the vehicle may run trouble free for a long period, until, with the advance of years, breakdowns become more frequent.

An interesting application of the 'bathtub' curve is in consideration of population dynamics. One factor affecting birth rate in a population is the average size of a family. In simple terms, more than two children reaching maturity constitutes an increase in population over one generation, so family size makes a critical contribution to the balance of birth and death rates. Very poor people are likely to have large families as their form of wealth generation, to provide more potential breadwinners, and to ensure being looked after in old age. As income per capita increases, the desired family size drops dramatically, until reaching a

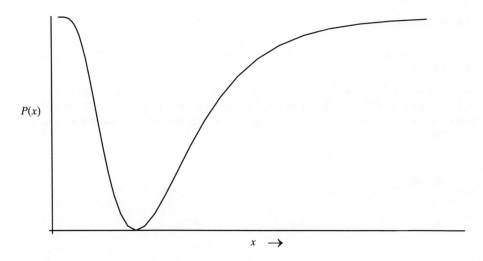

Figure 3.15   Inverted log-normal distribution.

minimum at levels of income enjoyed by the majority in developed countries. As income per capita rises further, the economic constraints felt by most people are less apparent, and there may be a tendency for family size to increase, albeit slowly, with increasing income.

3.3.3 Time Advance

By virtue of the fact that discrete models are characterized by only changing state at specific event times, discrete simulation time advance takes place in a series of jumps from one event to the next in chronological order. This being the case, it is only necessary to consult the event queue at any point, to ascertain the time of the next event to be processed, and to advance time to this point.

For systems that work simply by manipulating an event queue (event-oriented simulation packages), this approach works well, for an event is (usually) only placed in the event queue if it has been scheduled in advance, and therefore bound to take place when the time has been reached. Activity-based modelling on the other hand is a little more speculative, in that it relies on the concept of the *activity scan*. There will probably be an event queue in which all possible events are stored, particularly relating to the start of activities. However, just because an event is in the queue does not mean that it can take place. Activity modelling requires all the required entities to be in place before an activity can start. Activity scanning therefore is considered by some to be wasteful, in that much work is done without necessarily being able to advance time.

3.4 REFERENCES

The following texts are recommended for further reading on the subject of discrete simulation.

FISHMAN, G S (1978) *Principles of discrete event simulation*, New York: John Wiley.

NEELAMKAVIL, F (1987) *Computer simulation and modelling*, Chichester, UK: John Wiley.

PIDD, MICHAEL (1989) *Computer modelling for discrete simulation*, Chichester, UK: John Wiley.

4

Analog and Hybrid Computation

4.1 INTRODUCTION TO ANALOG COMPUTATION

After reading in Section 2.2 that the analog computer, in terms of present-day simulation practice, has largely been consigned to history, you may be wondering what to expect in a chapter on analog and hybrid computation. There are three main reasons for including these topics in the book:

- The analog computer was the forerunner in simulation hardware, being used for solving differential and algebraic equations before digital techniques were developed. Digital simulation software tended to follow the 'analog-emulator' route, for commercial reasons, and because numerical integration methods were generally oriented towards single-stage integration of ordinary differential equations.
- In spite of the demise of the analog computer, there are lessons to be learned from its use as a simulation tool. Many of the techniques of analog computation have their counterparts in digital simulation, and attention to some of these may help to alleviate implementation problems with digital machines.
- The almost universal adoption of the graphical user interface (GUI) has led to the development of graphical model input software based largely on analog computing block-oriented concepts.

In order to be able to appreciate the contribution that analog computation has made to continuous system simulation, it is necessary for us to understand a little about how an analog computer works in terms of producing solutions to the differential and algebraic equations that make up a simulation model. We should remember that the one very important fact about an analog computer is its ability to carry out the operation of integration, and, if necessary, to do it very quickly. To understand how this comes about, we need to know a little about the *operational amplifier* and its role in providing us with the building blocks of an analog computer model implementation.

4.1.1 The Operational Amplifier

Figure 4.1 illustrates a simplified circuit diagram for a high-gain DC amplifier with associated circuitry, the whole of which constitutes the so-called operational amplifier. Without going into details of the electronics, the high-gain amplifier (shown as the symbol shaped like a pie-slice) has a high negative amplification factor $(-A)$ which of itself is useless for computational purposes. With the feedback impedance Z_f and the input impedance Z_{in} connected in the manner shown through a summing junction (the $\oplus$ symbol in the diagram), the circuitry is now in a position to give us a relationship between input and output voltages. The impedances are defined by the simple Ohm's Law relationship

$$Z = e/i \tag{4.1}$$

where e is the potential difference or voltage drop across the impedance and i is the current flowing through it. The DC operational amplifier is designed so that:

- The output voltage e_o is related to the summing junction voltage e_b by the gain of the amplifier (i.e. $e_o = -A \cdot e_b$) within the reference range of the computer.
- The amplifier draws negligible current ($i_b \approx 10^{-9}$ amps).
- The gain of the amplifier is very high ($A \approx 10^8$). From this it can be seen that the effective summing junction voltage e_b is practically zero.

Applying Kirchhoff's Law to the summing junction:

$$i_b = i_{in} + i_f \tag{4.2}$$

or from Ohm's Law:

$$i_b = \frac{e_{in} - e_b}{Z_{in}} + \frac{e_o - e_b}{Z_f} \tag{4.3}$$

Since i_b is very small, it is assumed to be zero. Replacing e_b by $e_o/-A$, we obtain

$$\frac{e_{in}}{Z_{in}} + \frac{e_o}{A \cdot Z_{in}} = \frac{-e_o}{A \cdot Z_f} - \frac{e_o}{Z_f} \tag{4.4}$$

$$\therefore e_o = \frac{-(Z_f/Z_{in}) e_{in}}{1 + (1/A)[(Z_f/Z_{in}) + 1]} \tag{4.5}$$

but Z_f/Z_{in} is usually less than about 50 and $A \gg 1$,

$$\therefore e_o = -\frac{Z_f}{Z_{in}} e_{in} \tag{4.6}$$

Note the change of sign that occurs within an operational amplifier. We now have a linear relationship between the input and output voltages, in terms of the ratio of the two impedances involved.

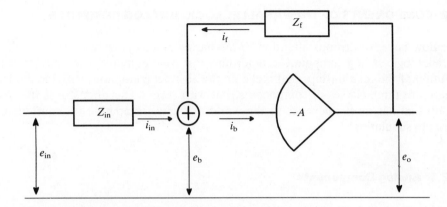

Figure 4.1 Basic electronic circuitry for operational amplifier.

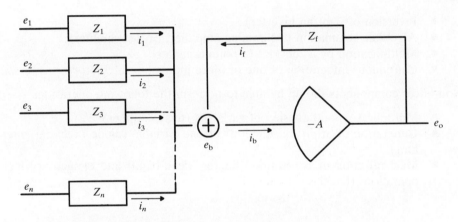

Figure 4.2 Operational amplifier with multiple inputs.

If we extend this network to include several inputs as shown in Figure 4.2, then applying Kirchhoff's and Ohm's Laws in the same way as above (remembering that $i_b \cong 0$ and $e_b \cong 0$), we may obtain the extended relationship

$$e_o = -\left[\frac{Z_f}{Z_1}e_1 + \frac{Z_f}{Z_2}e_2 + \frac{Z_f}{Z_3}e_3 + \ldots + \frac{Z_f}{Z_n}e_n\right] \qquad (4.7)$$

Equations (4.6) and (4.7) are the basis for all the mathematical relationships capable of being represented on an analog computer.

4.2 COMPONENTS OF THE PARALLEL LOGIC ANALOG COMPUTER

We now have to turn our attention to the hardware components of the so-called parallel logic analog computer, to determine just what equipment we need for the solution of the relationships that make up the abstract continuous-variable model. There are three classes of components, but with certain exceptions, it is the first group that has most relevance to the world of digital computing for continuous-variable simulation.

4.2.1 Analog Components

The pure analog components are those which have variable voltages as inputs and outputs. In order to be of any real use, we should be able to represent both linear and non-linear relationships. Linear components should be able to perform the operations of:

- Inversion of sign (an inverter)
- Algebraic summation (a summing amplifier, or just summer)
- Multiplication by a constant (a potentiometer)
- Continuous integration of one or more inputs (an integrator).

Non-linear components should be able to perform the following operations:

- Multiplication and division of variables (the multiplier)
- Generation of arbitrary functions of one variable (diode function generator)
- Mechanization of constraints (limiter, dead-band) and elementary logic operations (backlash, hysteresis).

4.2.1.1 Linear analog computer components

Let us consider first the linear component operations. In order to perform sign inversion, we need to consider how this is done in terms of equation (4.6). Quite simply, if the ratio Z_f/Z_{in} is set to unity, then equation (4.6) becomes

$$e_o = -e_{in} \tag{4.8}$$

and we have straightforward sign inversion. The time has come to be more specific about the nature of the impedances that would be needed in the operational amplifier for it to perform the function of inversion. If we replace Z_f by a resistor R_f and Z_{in} by R_{in}, such that $R_f = R_{in}$, then we shall achieve the desired result. Figure 4.3(a) shows the changes to the circuit. Figure 4.3(b) shows the programming symbol commonly used for the inverter. Programming, in the context of the analog computer, constitutes the preparation of a complete

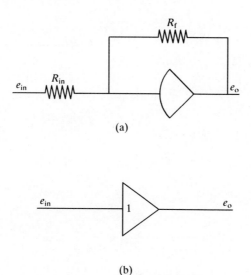

(a)

(b)

Figure 4.3   (a) Operational amplifier set up as inverter ($R_f = R_{in}$); (b) programming symbol for inverter.

symbolic circuit diagram for the model, and 'patching', that is making the appropriate machine connections between components.

If we make the same type of substitution for the impedances in the multiple input amplifier in Figure 4.2, the outcome is the summing amplifier. If R_f replaces Z_f and $R_1, R_2, R_3, \ldots, R_n$ replace $Z_1, Z_2, Z_3, \ldots, Z_n$ respectively, then we have the circuitry shown in Figure 4.4(a) and the output voltage is given by

$$e_o = -\left[\frac{R_f}{R_1} e_1 + \frac{R_f}{R_2} e_2 + \frac{R_f}{R_3} e_3 + \ldots + \frac{R_f}{R_n} e_n \right] \tag{4.9}$$

$$\therefore e_0 = -[G_1 e_1 + G_2 e_2 + G_3 e_3 + \ldots + G_n e_n] \tag{4.10}$$

Figure 4.4(b) shows the programming symbol for a multiple input summer. The fixed resistances available on an analog computer are of the order of 1 MΩ and 100 kΩ for transistorized or integrated circuit machines, so the substitution of values of this order in equation (4.9) would give us a rather limited range of coefficient values for any linear relationships we wished to implement. We may get around this problem by use of the potentiometer to give us 'infinite' variability over a range of coefficient values.

Figure 4.5(a) shows a simple diagram of the circuitry, while Figure 4.5(b) shows the programming symbol. A potentiometer is, in essence, a fixed resistor (usually wire wound) across the surface of which moves a 'wiper' arm. In analog computing work, it is used predominantly in 'grounded' or earthed mode, as shown in the diagram.

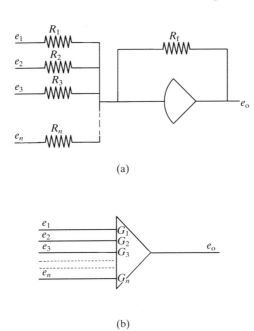

(a)

(b)

Figure 4.4   (a) Operational amplifier set up as summer; (b) programming symbol for summer.

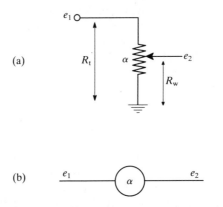

Figure 4.5   (a) Potentiometer circuitry details; (b) programming symbol for potentiometer.

There is a mechanical ratio for the potentiometer coefficient which is the ratio of the resistance between the wiper and ground, R_w and the total resistance, R_t. In the unloaded state (i.e. when no current passes along the wiper arm), the electrical ratio e_1/e_2 (known as α) is equal to the mechanical ratio R_w/R_t. However, when the system is loaded (as is the case when connected to an

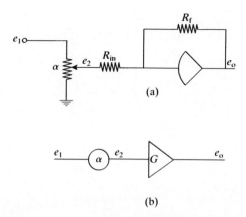

(a)

(b)

Figure 4.6 (a) Potentiometer input to single-input amplifier; (b) programming symbol for above circuit.

amplifier), α is not equal to the mechanical ratio R_w/R_t. The desired value for the coefficient α is set only after the amplifier connections have been made for the problem in hand. By the nature of the potentiometer, α is such that $0 \leqslant \alpha \leqslant 1$, but the value it can have is variable within that range. Figure 4.6(a) shows the circuitry involved in the connection of a potentiometer in the input path of a single-input amplifier. The ratio R_f/R_{in} for the operational amplifier resistances is known as the *input gain*, G (not to be confused with the high gain A that we came across earlier). Figure 4.6(b) shows the programming symbols for the potentio-meter input. Note the insertion of the gain G in the amplifier symbol. Using equation (4.6) we may write

$$e_o = -\frac{R_f}{R_{in}} e_2 \tag{4.11}$$

but

$$e_2 = \alpha.e_1 \tag{4.12}$$

$$\therefore e_o = -\alpha \frac{R_f}{R_{in}} e_1 = -\alpha.G.e_1 \tag{4.13}$$

We may extend this to the multiple input case. If the input resistors of Figure 4.4(a) are each preceded by a potentiometer, then the circuit diagram becomes that shown in Figure 4.7(a). Figure 4.7(b) shows the programming symbols used to represent the same set-up.

In this particular case, the output e_o is given by

$$e_o = -\left[\frac{\alpha_1 R_f}{R_1} e_1 + \frac{\alpha_2 R_f}{R_2} e_2 + \frac{\alpha_3 R_f}{R_3} e_3 + \ldots + \frac{\alpha_n R_f}{R_n} e_n \right] \tag{4.14}$$

$$\therefore e_o = -[\alpha_1 G_1 e_1 + \alpha_2 G_2 e_2 + \alpha_3 G_3 e_3 + \ldots + \alpha_n G_n e_n] \tag{4.15}$$

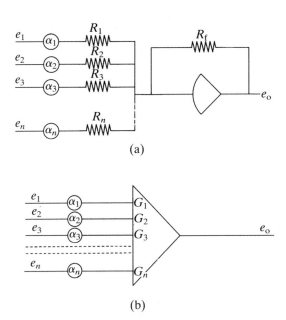

(a)

(b)

Figure 4.7   (a) Potentiometer input to multiple input amplifier; (b) programming
symbols for above circuit.

This gives us a considerably greater range of coefficients for the implementation
of linear algebraic equations on an analog computer. However, the inclusion of
potentiometer coefficients attenuates coefficient values in the input path. Inclusion of a single potentiometer in the feedback path of the amplifier (between the
output and the feedback resistor) would have the effect of dividing the whole of
equations (4.14) and (4.15) by the coefficient α_f which would enable higher
coefficient values to be obtained in linear equations.

In the diagram of Figure 4.3(a), if the resistor R_f is replaced by a capacitor C,
then we obtain the circuit shown in Figure 4.8(a). For a capacitor C, the
relationship between current and voltage is

$$i = C \, \frac{de}{dt} \tag{4.16}$$

where e is the voltage drop across the capacitor. Therefore, for the feedback
capacitor in Figure 4.8(a), the relationship is

$$i_f = C \, \frac{d}{dt} (e_o - e_b) \tag{4.17}$$

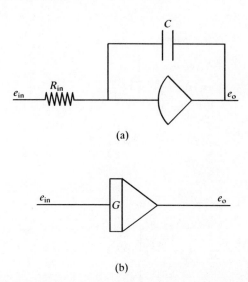

(a)

(b)

Figure 4.8 (a) Circuitry details for single-input integrator; (b) programming symbol for above circuit.

Since $i_b \cong 0$ and $e_b \cong 0$, we have the following relationship for current at the summing junction:

$$i_{in} + i_f = 0 \quad \text{(see Figure 4.1)} \tag{4.18}$$

$$\therefore \quad \frac{e_{in}}{R_{in}} + C\frac{de_o}{dt} = 0 \tag{4.19}$$

$$\therefore e_o = -\frac{1}{R_{in}C}\int_0^T e_{in}\,dt \tag{4.20}$$

This gives us the equation for an integrator with one input. On the assumption that there is no initial charge on C, the initial condition, or constant of integration, is zero. Figure 4.8(b) shows the programming symbol for this integrator. The overall gain for this integrator is given by

$$G = \frac{1}{R_{in}C} \tag{4.21}$$

$$= 1 \quad \text{if } R_{in} = 1\,\text{megohm (M}\Omega) \text{ and } C = 1\,\text{microfarad (}\mu\text{F)}$$

The term 'overall' is used here deliberately, for if we now replace the feedback resistor R_f with a capacitor C in Figure 4.4(a) we obtain the circuit diagram of

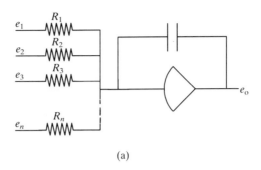

(a)

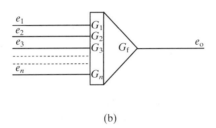

(b)

Figure 4.9 (a) Circuitry details for multiple input integrator; (b) programming
symbol for above circuit.

Figure 4.9(a). In the same way as was done for equations (4.16) to (4.20), it may
be shown that the output voltage of this circuit, e_o, is given by

$$e_o = -\frac{1}{C} \int_0^T \left[\frac{e_1}{R_1} + \frac{e_2}{R_2} + \frac{e_3}{R_3} + \ldots + \frac{e_n}{R_n} \right] dt \qquad (4.22)$$

The programming symbol for a multiple input integrator is shown in Figure
4.9(b). In this figure,

$$G_1 = 1/R_1, \; G_2 = 1/R_2, \; G_3 = 1/R_3, \; G_n = 1/R_n \text{ and } G_f = 1/C$$

The capacitor, being common to all terms in equation (4.22), is taken outside the
integral to form an overall integrator feedback gain that depends on the capacitor
value being used ($G_f = 1$ for $C = 1 \, \mu\text{F}$). We can now integrate a sum of inputs,
each term being multiplied by a constant coefficient. The use of potentiometers in
the input paths gives us further flexibility for setting coefficient values without
being restrained by fixed resistor/capacitor values. This covers the range of linear
operations available to us on an analog computer.

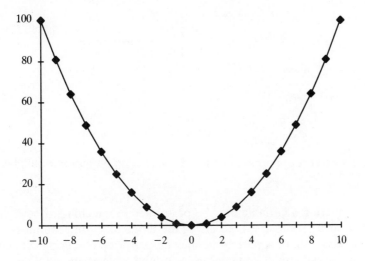

Figure 4.10   Straight-line segment approximation for $y = x^2$.

4.2.1.2 Non-linear components

In terms of relevance to digital computing for simulation, the arbitrary function generator (AFG) probably has the closest connection. On an analog computer, this component is one of the worst to set up, so much so that later machines had detachable AFGs which the simulationists of the day guarded with their lives. Once a function had been programmed into one of these devices, a simulationist was understandably very reluctant to have to repeat the exercise. AFGs required information in the graphical form $y = f(x)$. A curve giving a graphical representation of the function was first drawn out on graph paper. This graph was inserted into an X–Y plotter, and a series of straight-line segments were programmed via potentiometers and diodes, to give a 'best' fit to the curve on the graph paper. In some cases, this exercise could take all day! Figure 4.10 shows a straight-line segment plot for the function $y = x^2$. This leads us on to the next piece of non-linear equipment – the multiplier.

Although there are several analog computer devices that yield the product of two analog variables, the best-known one (and one of the most reliable) is the *quarter-squares multiplier*. This operates according to the identity

$$\frac{1}{4}[(x + y)^2 - (x - y)^2] = xy \tag{4.23}$$

It consists of two high-quality hard-wired diode function generators (to perform the operation of squaring) and associated amplifiers (for addition, subtraction and multiplication by 0.25). The programming symbol for a multiplier is shown in

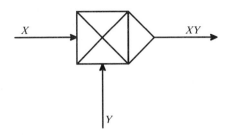

Figure 4.11   Programming symbol for an analog computer multiplier.

Figure 4.11. Squaring a variable may be performed by patching to both inputs of the multiplier, or, in some cases, accessing the diode function generators directly. Division and extraction of square roots may be performed by using the multiplier in feedback in special circuit configurations around a high-gain amplifier.

The remaining non-linear components in an analog computer may or may not be provided as such. Many machines provided a number of component parts, such as diodes and high-gain amplifiers, from which system non-linearities could be built up. These include limiter functions, hysteresis or backlash, absolute value and dead-zone. One important implementation that is worthy of mention is the ability to achieve position and rate limiting in the case of an integrator. Problems in aircraft dynamics, among others, often require the position limiting of an integrator, in such a way that the rate becomes zero at the integration limit. This can be done by diode feedback from the integrator itself, and from the amplifier producing the rate input to the integrator (Fifer, 1961).

4.2.2 Logic Components

The parallel logic analog computer is, as its name suggests, an analog computer that is provided with a complement of logic elements such as *AND* gates, *OR* gates, monostables and bistables, from which various logic operations could be programmed to control the action of the analog computer in carrying out simulation runs. In some problems where decision-making assumes a degree of importance in controlling a sequence of repetitive-operation simulation runs, a knowledge of Boolean and other logic algebras is useful to minimize the use of what is always a limited number of logic components available on a machine. The digital computer is far better for the implementation of complex logic, so the analog computer does not have very much to teach us in this direction. What is useful, however, is to look at the end product of any logic implementation, in terms of run control of the analog computer. In this respect, there is much to be gained by looking at the rules under which analog computers operate. In order to be able to communicate with the analog components, we need what are called *interface* components. These work in two ways. Either they accept analog signals and generate logic signals, or they accept logic signals upon which their subsequent analog output behaviour depends.

4.2.3 Interface Components

The first case above is exemplified by the *analog comparator* which generates a logic signal dependent upon the algebraic sum of two analog signals. The digital counterpart of this is the conditional *IF* type of construction which determines one or another course of action to be followed, depending on the logical truth of a relational statement.

More important from our point of view is the second type of interface component that accepts a logic signal and determines its analog behaviour accordingly. The D/A switch (a logic-operated ON/OFF switch in an analog circuit path), the relay (a similar type of device) and the track/hold unit are examples of this kind of operation. By far the most important interface unit, however, is the integrator itself, which we have so far seen only as an analog component.

Consider the solution of a first-order ordinary differential equation

$$\frac{dx}{dt} = f(x, y) \tag{4.24}$$

The analog computer solves such an equation by carrying out the inverse operation, that of integration, according to the equation

$$x = \int_0^T f(x, y)\, dt + x_0 \tag{4.25}$$

T is the period over which the integration process is carried out, and x_0 is the constant of integration, or more properly in terms of producing a solution for the definite integral, the *initial value* or *initial condition* which characterizes the form of the solution obtained. In order to produce such a solution, the analog integrator must operate in two modes. The first is the setting up of an initial value for the solution variable, and the second is the integration of the sum of inputs away from that initial condition. This bears a resemblance to aspects of simulation structure outlined in Section 2.4.3, so it is worth exploring this connection further in the case of the analog computer.

4.3 SIMULATION STRUCTURE FOR ANALOG COMPUTATION

Figure 4.12(a) shows a simplified circuit diagram incorporating the logic-controlled switches necessary for integrator mode control. C and R are two logic

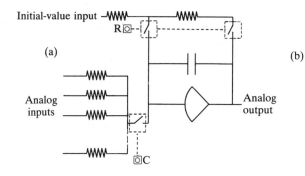

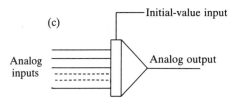

Mode	C	R
Not allowed	1	1
Compute	1	0
Reset	0	1
Hold	0	0

Figure 4.12 (a) Circuit diagram for integrator mode control; (b) truth table for integrator modes of operation; (c) programming symbol for integrator showing IC input.

mode control inputs. Although, strictly speaking, we need only one to provide the two operating modes described above, there is a third mode known as 'hold' which is available on most machines. We shall examine the uses to which this may be put, in due course. The 'truth' table in Figure 4.12(b) indicates the logic signals needed to operate the integrator in each of the three modes allowed. In the circuit, a logic '1' closes the appropriate switch(es), so we can examine what happens to the integrator in each mode of operation.

- In *Reset* mode, the C switch is open and the R switches are closed. The normal analog inputs are disconnected from the summing junction, and instead the initial value input is applied across the network at the top of the diagram. Low-value input and feedback resistors, both of the same value, together with the capacitor, make up a very fast-acting circuit with what is known as *first-order lag* operation. Upon sensing the presence of an input voltage corresponding to the initial value, the capacitor charges up exponentially, until the analog output matches this value, but with change of sign. The integrator has now been 'reset' to initial conditions, ready for the compute mode to begin. This corresponds to the *initialization* stage of simulation structure as outlined in Section 2.4.3.

- In *Compute* mode, the R switches are opened and the C switch closed to apply the analog inputs. The feedback capacitor is now charged to the initial conditions value, and proceeds to integrate the sum of inputs away from this value, thus producing the solution to an equation of the form shown in equation (4.25). This is the *dynamic* stage of the analog computer simulation.

- The remaining mode of operation open to us from the truth table is the so-called *Hold*. When all inputs are removed from the summing junction of the amplifier, the capacitor charge and hence the analog output remain constant. This feature is very useful to apply at the end of a simulation run, when, for example, the final values of the simulation transients may determine the initial values for the next run. Keeping the final values constant for a short period enables the calculation of new initial values to be done properly before *Reset* mode is once again entered. This is the analog computer equivalent of the *termination* stage.

From the above, we can see that the interface operation of the integrator under logic action is responsible for the way that all the analog integrators behave, and hence how a complete simulation problem will behave when implemented on an analog machine. We can now take this process a stage further in exploring the process known as repetitive operation which, by reason of its high-speed solution capability, the analog computer does very well. There is a fundamental problem common to all analog machines, namely that of range of operation. DC amplifiers are designed to operate within a specific voltage range, in order to retain their linear characteristics. On the older machines using valve (vacuum-tube) circuitry, the range was ±100 volts, but semiconductor and integrated circuit machines run at ±10 volts. The variables of a simulation model can acquire a large range of values over the course of a simulation run, and the need therefore arises in analog implementation to scale the range of operation of each physical variable to correspond with the voltage range of the machine. This can be a very laborious process, especially for large models, and if the anticipated operating ranges of variables turn out to be wrong, the exercise may need to be repeated several times until all operating constraints are met.

Amplitude scaling, as it is called, is not a problem that one normally associates with digital computation. The range of values available through the medium of a 32 bit digital 'word' is usually more than enough for most applications, so it would probably be safe to say that amplitude scaling is not going to be relevant computationally to digital computer simulation. If, however, we look back in Section 2.6 at the world dynamics simulation graphical output in Figure 2.4, we can see a good example of amplitude scaling applied to the output of transients that have been obtained from a digital computer implementation of a simulation. This was necessitated by the very different magnitudes of the variables being plotted, and the desire to include all transients on one plot without a multitude of *Y*-axis scales being present. 'Normalization' of the transients to their 'maximum' values gave a single *Y*-axis range from 0 to about 1.0–1.5, suitable for all the transients, thus giving all possible space to the plots themselves.

The other form of scaling frequently applied to analog computer simulation is known as *time scaling*. The ability to operate at solution speeds very much faster than real time gave the analog computer its ascendancy for many years, even when digital machines were being applied to simulation problems. In order to take advantage of this high solution speed, one needs to introduce a timescale factor into the physical equations for the integrators. This has the effect of reducing the size of capacitors required in integrator feedback, and hence speeding up the solutions obtained. This does have some relevance to the digital scene, for many problems solved digitally will run faster than real time, or can be made to do so.

It is not necessary for us to deal with the actual mechanics of scaling, either for amplitude or for time. It is, however, useful for us to look at other more general aspects of model implementation on an analog machine to see if there are lessons to be learned for digital computation. Two examples are given. One is a straightforward application to the solution of classical second-order differential equations, while the second illustrates the use of an analog computer for iterative solution, that is repetitive operation.

4.3.1 Model Implementation and Initialization

4.3.1.1 Second-order differential equations

To illustrate the application of analog computation to the solution of differential equations, we shall cover the steps necessary for derivation of the so-called 'machine' equations for two well-known second-order systems. The procedures in each case will enable us to draw the patching diagram for the problems, without undertaking any amplitude or time scaling.

An undamped simple harmonic motion is described by the equation

$$\frac{\mathrm{d}^2 x}{\mathrm{d}t^2} + 4x = 0 \tag{4.26}$$

with initial value x_0 ($\dot{x}_0 = 0$).

For analog computer solution, we need to decompose equation (4.26) into two first-order differential equations. We may write equation (4.26) as

$$\frac{\mathrm{d}\dot{x}}{\mathrm{d}t} = -4x \tag{4.27}$$

where

$$\frac{\mathrm{d}x}{\mathrm{d}t} = \dot{x} \tag{4.28}$$

From an analog programming point of view, it is more useful to express equations (4.27) and (4.28) as integral equations thus:

$$\dot{x} = \int_0^T -4x\,dt + \dot{x}_0 \tag{4.29}$$

$$x = \int_0^T \dot{x}\,dt + x_0 \tag{4.30}$$

The reason for this is that, during the operation of amplitude scaling, the scaling operations are automatically applied to the initial values as and when required. We now have to take into account the sign inversion that occurs with analog computer amplifiers. With reference to Figure 4.13(a), if we choose to solve for $+x$ (so that we see the solution the right side up), then the derivative $\dot{x}$ in equation (4.30) (input to integrator 1) has to be negative. Since we have inverted the output of integrator 2, the input will have the sign of the derivative in

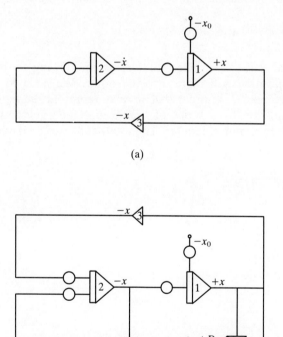

(a)

(b)

Figure 4.13 (a) Analog patching diagram for undamped second-order system; (b) analog patching diagram for Van der Pol's equation.

equation (4.29) without any sign change. That means that the input has to be $-x$ (in keeping with equation (4.29)), so that an inverter is needed in the circuit (amplifier 3). Algebraically, the process of sign inversion is taken into account by writing the equations as

$$-(-\dot{x}) = \int_0^T -4x\,\mathrm{d}t \tag{4.31}$$

$$-(+x) = \int_0^T -\dot{x}\,\mathrm{d}t - x_0 \tag{4.32}$$

thus necessitating an inverter to give

$$-(-x) = +x \tag{4.33}$$

The minus signs preceding the solution variable in each case represent algebraically the sign inversion taking place in the amplifier.

The second equation is an extension of the first, being the well-known Van der Pol equation for non-linear limit-cycling oscillatory behaviour:

$$\frac{\mathrm{d}^2 x}{\mathrm{d}t^2} - k(1-x^2)\frac{\mathrm{d}x}{\mathrm{d}t} + 4x = 0 \tag{4.34}$$

The same initial value x_0 applies to this system. Owing to the non-linearity present, decomposition is a more involved affair. We need to create auxiliary variables A, B, C and D, in order that we may represent the non-linearity in the Van der Pol equation by a series of simple machine equations which may be implemented on an analog computer. The total set of unsigned machine equations is given as

$$\dot{x} = \int_0^T (A - 4x)\,\mathrm{d}t \tag{4.35}$$

$$x = \int_0^T \dot{x}\,\mathrm{d}t + x_0 \tag{4.36}$$

$$A = k.B \tag{4.37}$$

$$B = C.\dot{x} \tag{4.38}$$

$$C = 1 - D \tag{4.39}$$

$$D = x^2 \tag{4.40}$$

The patching diagram is shown in Figure 4.13(b). Note that the extra components needed are in line with the number of equations above. Strictly speaking, we do not need equation (4.37), but if we were carrying out a simulation experiment in which we wished to observe the behaviour of the system for various values of the parameter k, then it would be useful to have this value explicitly set by means of a single potentiometer (shown labelled k before amplifier 4). It is left to the reader to produce the signed machine equations for the system. (It should be assumed that multipliers do not produce sign inversion, i.e. they operate strictly algebraically.)

4.3.1.2 Two-point boundary-value problem (bending beam)

In the previous examples, initial values were explicitly stated for each of the systems. In this example, let us consider a beam bridging a gap between two walls, and with the ends embedded in the brickwork. The beam will deflect under its own weight. The equation relating the deflection y with distance x along the beam from one end is

$$EI \frac{d^4 y}{dx^4} = -W \tag{4.41}$$

where

  E is Young's modulus of elasticity
  I is the second moment of area of the beam cross-section
  y is the deflection
  x is the distance from one end of the beam (total length L)
  W is the weight of the beam per unit length

For analog computer implementation, equation (4.41) has to be decomposed into four first-order equations, and the independent variable t on the analog computer which normally represents simulation time now has to represent the independent variable x which is distance along the beam. Re-writing equation (4.41) as

$$\frac{d^4 y}{dx^4} = -\frac{W}{EI} \tag{4.42}$$

we may then define four auxiliary variables

$$\dddot{y} = y_4, \quad \ddot{y} = y_3, \quad \dot{y} = y_2, \quad y = y_1$$

and write down the four first-order differential equations

$$\frac{dy_4}{dx} = -\frac{W}{EI} \tag{4.43}$$

$$\frac{dy_3}{dx} = y_4 \tag{4.44}$$

$$\frac{dy_2}{dx} = y_3 \tag{4.45}$$

$$\frac{dy_1}{dx} = y_2 \tag{4.46}$$

In order to include initial values, it is more expedient to re-write these as integral equations

$$y_4 = \int_0^L -\frac{W}{EI}\,dx + y_{4_0} \tag{4.47}$$

$$y_3 = \int_0^L y_4\,dx + y_{3_0} \tag{4.48}$$

$$y_2 = \int_0^L y_3\,dx + y_{2_0} \tag{4.49}$$

$$y_1 = \int_0^L y_2\,dx + y_{1_0} \tag{4.50}$$

In order to solve these equations for y_1 (y), initial values are needed for y_4 ($\ddot{y}$), y_3 ($\dddot{y}$), y_2 ($\dot{y}$) and y_1 (y). Since the beam is embedded, y_{1_0} (y_0) and y_{2_0} ($\dot{y}_0$) are known because

$$y_1\ (y) = 0 \text{ at } x = 0 \text{ and at } x = L \tag{4.51}$$

$$y_2\ (\dot{y}) = 0 \text{ at } x = 0 \text{ and at } x = L \tag{4.52}$$

where L is the total length of the beam. y_{4_0} ($\ddot{y}_0$) and y_{3_0} ($\dddot{y}_0$) are not known and must be found either by trial and error, or by the application of an iterative method to converge the solution towards the correct values. Figure 4.14 shows qualitatively what the solution should look like when it has converged. Because of embedding in the wall at each end, the deflection and the beam slope are both zero at both ends of the beam. The problem may be stated as follows: What values of y_{4_0} and y_{3_0} do we need in order to achieve values $y_{1_L} = 0$ and $y_{2_L} = 0$?

We could set up the problem on an analog computer, have two potentiometers patched to give us initial values on y_3 and y_4, and spend much time trying to obtain a combination of settings that will yield zero final values for y_1 and y_2. However, it is useful to take advantage of the repetitive operation capabilities of the analog computer to obtain automatic convergence of the solution. This is achieved in the following way.

The integrators labelled 1, 2, 3, 4 in the analog patching diagram in Figure 4.15(a) constitute the analog implementation of the model equations, (4.47) to (4.50). Integrators 1 and 2 require non-zero initial conditions to be determined in such a way that, at the end of the Compute stage (at a point where simulation time t is equivalent to the length of the beam L), the final values of y_1 and y_2 are both zero. To achieve this, connect two integrators 5 and 6 into the circuit as shown, together with potentiometers and inverters. These integrators will

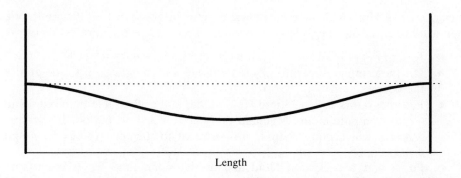

Figure 4.14   Bending beam embedded in a wall at each end.

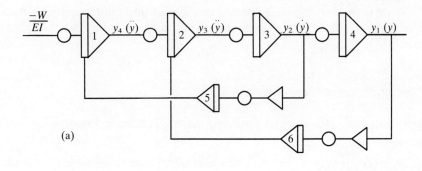

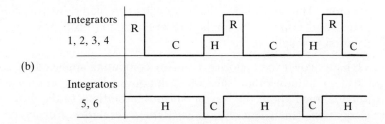

Figure 4.15   (a) Analog patching diagram for bending beam problem; (b) integrator mode timing diagram for bending beam problem.

normally be in Hold with short Compute periods in between (see diagram). We then need to carry out the following iterative procedure (see Figure 4.15(b)):

- Reset integrators 1, 2, 3, 4 (all with zero initial values to start).
- Put these integrators into Compute mode and integrate up to the end of the beam.
- Put integrators 1, 2, 3, 4 into Hold mode, and at the same time put 5 and 6 into Compute mode. Maintain this state for a short period. Integrators 5 and 6 will acquire outputs proportional to the final values on 3 and 4 respectively.
- Put 5 and 6 back into Hold and at the same time Reset integrators 1, 2, 3, 4 for another run (this time with 1 and 2 having non-zero initial values).

If this iterative sequence is repeated many times, the final values on 3 and 4 will converge to zero, at which point the outputs of 5 and 6 will become constant at the required initial values for 1 and 2. If the analog computer is speeded up so that a simulation run takes milliseconds, then the convergence is very rapid indeed. Mode control of the integrators is carried out through the implementation of a suitable logic circuit. The description of this circuitry is beyond the scope of this book.

4.3.2 Generating the Transients

We have dealt with two ways of obtaining analog computer solutions to sets of model equations:

- the *once-through* solution, when a straightforward initial-value problem has to be solved for a known set of parameters and initial conditions; and
- the *iterative* or *repetitive operation* technique, where convergence is required to a set of conditions that are not known at the start of the solution. This applies to boundary-value problems such as the bending beam, and also can be applied to optimization problems where a 'best' solution is sought.

These techniques cover the majority of analog computing applications, so we shall look no further at any other solution techniques. There are, however, some lessons to be learned about the way in which the solutions to the differential and algebraic equations (the *solution transients*) are generated and observed.

   Analog computer solution output is almost always of a graphical nature. Since the model variables are represented by machine voltages, we would expect to observe these through the medium of an X–Y plotter or a display oscilloscope. In the model development and testing phases, we would normally view an oscilloscope screen to obtain an instant response to any changes that are made. Since the analog computer is very much a 'hands-on' machine, this feedback is very valuable to enable the simulationist to acquire that necessary 'feel' for the model under development. For iterative solution, the same holds true, but for different reasons. Analog computer solution convergence often takes many hundreds of simulation runs to achieve, owing to the relatively unsophisticated convergence techniques being used, so any means of speeding this up is welcome. High-speed operation will establish the conditions for solution convergence. A single-pass slow-speed run can then be carried out with the required conditions in place. At this point, *X–Y* plotter output is the normal practice to enable hard-copy results to be obtained.

Analog computer graphical transient output may seem crude and inefficient. In simulation practice, this is not really the case. It is usually far easier to assess the correctness of results by examination of graphical output than it would be by looking at tabulated numerical printout. It is usually very easy to scan all the analog outputs of interest in order to assess the model performance, and this can be a very useful aid to model development and testing. The analog computer does lose out on the available precision of solution, but since most models contain inaccuracies in terms of model structure and data, too much precision is unnecessary. The best precision available from an analog computer is about 0.1%, but it is likely to be worse, depending on the nature and size of the model implementation.

4.3.3 At the End of a Run

In concluding our short look at the world of analog computing, we should take a look at what we would expect to see at the end of a simulation run, besides the solution transients drawn by an *X–Y* plotter.

This really depends on the nature of the solution technique applied to the problem. For a 'once-through' solution, we may not need to record any final values, since it is often the actual solution transients that are of most interest as a pointer to the dynamical behaviour of the model. Iterative solution methods will often require us to take note of the convergence conditions reached after repetitive operation of the machine has taken place over many solution runs. This is often done by reading the voltage outputs of certain amplifiers that have reached a steady state upon solution convergence. Observation of these outputs also gives an indication that the solution has indeed converged.

Armed with minimal knowledge of analog computation, we can now go on to look at the operation of the hybrid computer. This is the next phase of our movement towards digital computation as *the* medium for continuous-variable simulation.

4.4 INTRODUCTION TO HYBRID COMPUTERS

Without going into much detail about the operation and programming of analog computers, it may have become apparent to you that there are advantages and disadvantages of analog computation when compared with its digital counterpart for the purposes of continuous-variable simulation. The high speed of operation and 'hands-on' interactive nature of the analog computer are offset by the relative lack of working precision and the notorious programming difficulty associated especially with large problems. The digital computer scores by being relatively easy to program and debug, and by offering a high degree of precision.

The hybrid computer is the result of trying to achieve 'the best of both worlds' by having within a single unit both kinds of computer linked by an interface. With the advent of transistorized circuitry in the sixties, the reliability and speed of operation of various components increased to the point where the hybrid computer became a viable proposition. In particular, the evolution of fast reliable analog-to-digital and digital-to-analog converters facilitated the development of interface modules suitable for hybrid computation. At one stage during the era of the hybrid computer, it was very fashionable for academic institutions to have their own hybrid installations. Vast sums of money were spent in interface development for a diversity of configurations that seemed to be permutations of all the analog and digital equipment then available. There was also a considerable variety of commercially available hardware, but, in the main, there were two players who dominated the market in the hybrid computer stakes. The capital costs of equipment were immense, and running costs not much better. Although component reliability was good on the whole, the sheer complexity of the machines gave rise to high maintenance overheads.

Although the hybrid computer is little used at the present time, there are lessons to be learned from its application to simulation that may help us get the best from our use of digital computers. First, however, we should take a look at the way in which a hybrid computer functions, and the manner in which it was put to use for simulation purposes.

4.4.1 Hybrid Computer Architecture

Figure 4.16 shows a schematic diagram of the 'architecture' of a typical general-purpose hybrid computer. It consists of three major components:

- a *parallel logic analog computer* (i.e. an analog computer with a complement of patchable logic components for mode control operations),
- a *digital computer* with input/output ports to enable communication with
- an *interface*, which has all the hardware necessary to facilitate two-way communication between the analog and digital machines.

There are four main lines of communication across the interface, between the two computers. Two of these are for the exchange of information, and two are for control and notification of status:

- *Digital-to-analog word-to-scaled fraction transfer* by means of which one or more digital word values are read serially into a number of digital-to-

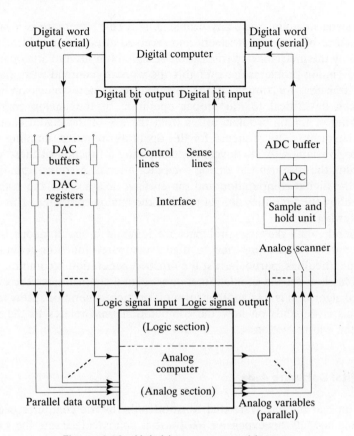

Figure 4.16 Hybrid computer architecture.

analog converter (DAC) buffers. When all required values are in place, they are transferred simultaneously into the same number of DAC registers, from which they are immediately 'seen' by the analog computer. This simultaneous transfer largely avoids what is known as a 'data-skewing' problem that would arise if the results were 'seen' piecemeal by the analog computer.

- *Analog-to-digital scaled fraction-to-word transfer* through which medium a series of analog scaled fraction values are set up and scanned in turn by an analog scanner, each to be held constant on a 'sample-and-hold' unit while an analog-to-digital converter (ADC) converts the scaled fraction value to a digital word representation. An ADC buffer holds the word until it is ready for transfer to the digital machine.
- *Control lines for digital-to-analog logic signal transfer*, whereby the digital computer sends a control bit signal for analog run control.
- *Sense lines for analog-to-digital logic signal transfer* by which the digital computer is informed of a change of status of the analog machine.

The logic signals do not cause any problems as a rule, for there is a one-to-one correspondence between what is sent and received at either end. Most problems are caused by the mechanics of translation of digital output into analog input, and vice versa. Timing problems are probably the worst to contend with, particularly for DAC transfer, for timing of the appearance of new values on the DAC registers can be critical to the proper operation of the analog computer. A different kind of timing problem arises from the use of the digital computer for function generation. The sequence for this operation requires sampling of one or more analog variables, A/D conversion, computing a function of these variables, and returning the result to the analog computer after D/A conversion. The serial nature of the digital computation and the ensuing delays cause the results to be 'out of date' by the time they are returned to the analog computer. This is known as 'time skewing'.

In general, 'data-skewing' problems are minimal if simultaneous transfer is arranged, but 'time skewing' due to digital computer function generation can cause some difficulties, particularly if the function generation is complex. This can have repercussions for the simulation of direct digital control, since the evaluation of a control algorithm is just another form of function generation. How much of a problem this is depends on how the computation time relates to the sampling period of the control action.

4.4.2 Digital Software Aids

After our look at the hardware that constitutes the hybrid computer, we need to consider just how all this expensive hardware is going to facilitate the solution of simulation problems. One question that immediately springs to mind is the degree to which the digital computer can assist in the setting up and checking out of the analog computer. There is also the question of analog program control which may best be handled by the digital machine. Let us consider these two aspects separately.

In a well-specified hybrid computer, the hardware is often in place to carry out a great many useful set-up and checking procedures under software control from the digital machine. For instance, the addition of a complement of *servo-set potentiometers* to the analog hardware, together with the availability of *digital–analog multipliers (DAMs)*, enables the automatic setting of potentiometer coefficients during problem set-up. Under software control too, the analog scanner–ADC system can read all analog amplifier outputs during what is called *static check mode*. This may be synonymous with Reset mode which we have seen before. In some machines, however, it requires the application of arbitrary values rather than true initial values from which all amplifier outputs are checked to establish the correct patching and setting up of the problem. Commercially available hybrid computers usually have interpretive software with which to specify the problem in digital mode. The amplifier outputs calculated from this operation can be compared with those actually obtained on the analog computer. Any discrepancies found can be used to track down analog set-up and patching errors. This automated set-up and checking facility is probably one of the better features of hybrid operation. We now need to examine run control to see if the same applies.

The running of an analog computer alone is, as we saw earlier, entirely under the control of a simulationist carrying out one run at a time and examining the results. Repetitive operation is carried out under the action of programmable logic on the machine. Far greater complexity of run control is possible on a hybrid computer, owing to the availability of the decision-making processes inherent in digital computer operation. Let us consider, by way of a simple example, the bending beam problem of Section 4.3.1.2, applied to a hybrid computer. The sequence of operating control steps might take the following path:

- Set up the first run with zero initial values on integrators 1 and 2 (see Figure 4.15(a)). These are set up on DAMs by the digital computer instead of coming from integrators 5 and 6 as shown in the diagram.
- A bit output from the digital computer puts the analog computer into Compute mode, via the programmable logic patched into the integrators.
- At the end of the run, the analog computer goes into Hold while the integrator outputs are converted and sent to the digital computer.
- The digital machine sets up new values (now non-zero) for the initial conditions on integrators 1 and 2, ready for another run.
- When the final values of integrators 3 and 4 are sufficiently close to zero, the sequence comes to an end.

Under digital control, use can be made of accelerated convergence techniques to lessen the number of analog computer runs needed to achieve the desired solution. In this way, therefore, the digital computer can help achieve a more efficient solution in terms of the number of runs required to reach the intended goal. This is a trivial example of such an application, but the principle holds good for more sophisticated problems where stand-alone analog solution is not possible.

4.4.3 Model Partitioning on a Hybrid Computer

The bending beam problem represents one example of the way in which a problem may be partitioned between the analog and digital computers of a hybrid machine. This represents an iterative mode of solution where analog and digital activity are sequential to each other, with results being passed back and forth. The other extreme exists where digital and analog computers function simultaneously, with timed transfer of results back and forth between the two machines.

The simulation of direct digital control is a very good application of the hybrid computer operating in this parallel mode. The analog computer produces the continuous solution of the model equations for the plant being controlled. The digital computer, on the other hand, handles the discrete sampling and calculation of a feedback control algorithm. Although the setting up and checking of this type of control problem is laborious, there are advantages that accrue from a hybrid computer solution, for instance:

- Hybrid computer simulation of computer control problems reproduces very closely the time- and data-skewing effects that will be experienced with real control hardware. It is possible to represent these to some degree by digital methods of simulation, as we shall see in later chapters, but other more subtle effects of data transfer may be more difficult to realize accurately within digital-only solutions.

- The use of hybrid computation will often highlight the *ill-conditioned* nature of a simulation problem. This is most likely due to the inherent electrical 'noise' present in the analog computer, which precludes the possibility of solution repeatability if the system model is very sensitive to slight changes. As we shall see, a digital-only solution is completely repeatable, so 'ill-conditioning' is not easily detectable unless special measures are taken.

A further example of problem partitioning in parallel on a hybrid computer is that in which the model itself is partitioned between the analog and digital machines. A very good example of this regime is a simulation model of a space re-entry vehicle (Electronic Associates Inc., 1965). This was a typical hybrid computer application during the 'space race' where the point dynamics of vehicle attitude and control were simulated on the analog computer, while the high-precision trajectory calculations were done on the digital computer. The fast reactions required of the attitude control system were well suited to analog implementation, while the relatively very slow movement of the vehicle along a space trajectory required the high precision that ensured accurate prediction of the point of landing, and where that was possible, accurate control of the vehicle to land in that area.

4.5 REFERENCES

ELECTRONIC ASSOCIATES INC. (1965) 'Hybrid simulation of a temperature rate flight control system for re-entry vehicles', Application Study No. 3.4.13h, Applications Reference Library, West Long Branch, NJ, Electronic Associates Inc.

FIFER, STANLEY (1961) *Analogue computation*, New York, McGraw-Hill, pp. 1267–74.

5

Digital Computation for Continuous Simulation

5.1 CHARACTERISTICS OF CONTINUOUS MODELS

The example problems presented in the previous chapter showed typically that we started with ordinary differential equations of order greater than one, and reduced them to first-order form by decomposition, in preparation for analog computer implementation. By and large, continuous models for digital computer implementation are treated in the same way, so that the differential equations are all reduced to first order. As we shall see in the next chapter, the standard way of handling the numerical integration of differential equations requires that they be first order. In this respect, there is no real difference between analog and digital computation.

We also had an example of a problem that featured algebraic equations to define auxiliary variables. Few real world models consist entirely of differential equations and the deficit between the number of variables in the model and the number of differential equations is made up by an assortment of possible relationships. These include algebraic equations (like those in the analog model), arbitrary functions (constructed as piecewise linear approximations to empirical curves), system non-linearities (such as limiters, dead-zones, hysteresis, backlash, etc.) and transfer functions (which are input/output relationships presented as ratios of polynomials). We can now begin to consider each of these in terms of their implementation within a digital computer simulation of a continuous system.

5.1.1 Differential Equations

In Section 4.2.3, we saw how the analog integrator solved differential equations written in the form of equation (4.24) or integral equations of the form shown in equation (4.25). As far as the analog computer is concerned, the inputs to the integrator (the derivative) appear on the right-hand side of the equation. Only the solution variable appears on the left. At present, we may assume that this is the case also with differential equations presented for digital computer solution. Some simulation packages require the equation to be written in a form akin to equation

(4.24), that is as differential equations with a term representing a first-order derivative on the left of the 'equals' sign. Other packages use the integral equation form of equation (4.25) which has the advantage that the initial value is usually part of the integral equation expression. High-order differential equations, therefore, have to be decomposed into sets of first-order equations in the same way as for analog computation. There is, however, no need to decompose complicated derivative expressions into simple component parts, as is the case with an analog machine.

5.1.2 Algebraic Equations

We now have to give a little thought to the part that algebraic equations play within a continuous-variable simulation, and the manner in which we should handle them. In the simple example afforded by the analog solution to the Van der Pol equation, we ended up with a number of simple component-based algebraic equations for each of which an output variable (the output of an amplifier) resulted from the analog implementation of a mathematical expression representing the inputs to the system. For the moment, therefore, let us assume that the same principle holds for digital computer implementation, namely that the result of evaluating an expression on the right-hand side of a computer coding statement is transferred to a single variable on the left-hand side (the solution variable).

5.1.3 Arbitrary Functions

The principles of piecewise linear arbitrary function generation have been adopted more or less intact for digital computer implementation. The only difference lies in the manner in which the function is mechanized. The analog computer AFG required a graphically aided set-up procedure to set the potentiometers and diode cut-off points to generate the voltage output appropriate to a given input. Digitally, a number of piecewise linear segments are set up, but this time as co-ordinate pairs. Function generation requires a choice of the segment appropriate to the input, and linear interpolation between the corresponding pairs of co-ordinate points. This is where the similarity ends. Digital AFGs are able to go much further than their analog counterparts. An analog AFG is one dimensional, that is it represents a function $y = f(x)$. Digital AFGs can represent functions in many dimensions, yielding representations of functions of the form $y = f(x_1, x_2, . . .)$. Such representations are, in essence, look-up tables, with linear interpolation applied between appropriate co-ordinate pairs within the table matrix.

5.1.4 Special Non-linear Functions

The analog computer diode function generators that yield many of the special non-linearities required in modelling work have their digital counterparts. However, where patching and setting up an analog limiter, dead-zone, or backlash function can be time consuming, the digital equivalents are relatively easy to program. Most of these are simple functions with the appropriate decision-making processes incorporated to effect correct functional behaviour. For example, one could write a simple FORTRAN function for a limiter as follows:

```
REAL FUNCTION LIMIT(HIGH, LOW, X)
REAL HIGH, LOW, X

LIMIT = X
IF(X.GT.HIGH)LIMIT = HIGH
IF(X.LT.LOW)LIMIT = LOW
RETURN
END
```

Other function non-linearities are also relatively easy to program compared with the analog equivalents.

5.1.5 Transfer Functions

Those of you who have – even briefly – come across some of the concepts of control engineering will probably have seen a transfer function. For those of you who have not, a transfer function is a relationship of one or more outputs to one or more inputs (not necessarily the same number). Let us consider a single-input–single-output (SISO) system. The input/output 'mapping' in this case takes the form of a ratio of two polynomials, for example

$$\frac{Y(s)}{X(s)} = G(s) = \frac{a_m s^m + a_{m-1} s^{m-1} + a_{m-2} s^{m-2} + \ldots + a_1 s + a_0}{b_n s^n + b_{n-1} s^{n-1} + b_{n-2} s^{n-2} + \ldots + b_1 s + b_0} \qquad (5.1)$$

If we wish to see the transfer function in the accustomed $y = f(x)$ form, we would write the equation as

$$Y(s) = G(s) \cdot X(s) \qquad (5.2)$$

We need to look at the properties of the transfer function and the way in which it is handled as part of a simulation model. First of all, in a real physical system model, $n \geq m$, which means that the order of the numerator polynomial is no greater than that of the denominator. Expressed qualitatively, this just means that free derivative action is seldom present in a model without being counteracted by some integral action.

It is quite common for modelling elements to be expressed in transfer function form, for it is a compact way of conveying much modelling information and complexity. A transfer function is a mathematical expression in terms of polynomials in s, the Laplace operator. This is a complex-variable domain operator which represents the time-domain operation of differentiation (i.e. loosely speaking, $sy \equiv dy/dt$ if initial conditions are assumed to be zero). If we rewrite equation (5.1) as

$$(b_n s^n + b_{n-1} s^{n-1} + b_{n-2} s^{n-2} + \ldots + b_1 s + b_0) Y$$

$$= (a_m s^m + a_{m-1} s^{m-1} + a_{m-2} s^{m-2} + \ldots + a_1 s + a_0) X \qquad (5.3)$$

and we extend the time-domain equivalence to include higher powers of s to give $s^2 y \equiv d^2 y/dt^2$, $s^3 y \equiv d^3 y/dt^3, \ldots, s^n y \equiv d^n y/dt^n$, then we may express equation (5.3) in time-domain form as

$$b_n \frac{d^n y}{dt^n} + b_{n-1} \frac{d^{n-1} y}{dt^{n-1}} + b_{n-2} \frac{d^{n-2} y}{dt^{n-2}} + \ldots + b_1 \frac{dy}{dt} + b_0 y$$

$$= a_m \frac{d^m x}{dt^m} + a_{m-1} \frac{d^{m-1} x}{dt^{m-1}} + a_{m-2} \frac{d^{m-2} x}{dt^{m-2}} + \ldots + a_1 \frac{dx}{dt} + a_0 x \qquad (5.4)$$

This is nothing more than a high-order linear differential equation with respect to time that describes the dynamical behaviour of a subsystem of the model. We have come across the concept of decomposition into first-order ordinary differential equations as one of the necessary steps for analog computer programming, and since the same model presentation philosophy is retained for digital computer solution, we need to carry out a similar decomposition.

5.2 MODEL MANIPULATION FOR GENERAL-PURPOSE IMPLEMENTATION

In order to solve high-order differential equations on an analog computer, we first of all carried out a decomposition to convert an nth order differential equation into n first-order ordinary differential equations. The examples were second- and fourth-order systems, but the principles are the same for any order of system. The decomposition is relatively easy for a system where all derivative terms are derivatives of the solution variable, in which case we may proceed in line with the example below. If we have the differential equation

$$b_n \frac{d^n y}{dt^n} + b_{n-1} \frac{d^{n-1} y}{dt^{n-1}} + \ldots + b_1 \frac{dy}{dt} + b_0 y = f(x) \qquad (5.5)$$

then we may solve it for the highest derivative of y with respect to t by re-arranging to give

$$b_n \frac{d^n y}{dt^n} = f(x) - b_{n-1} \frac{d^{n-1} y}{dt^{n-1}} - \ldots - b_1 \frac{dy}{dt} - b_0 y \qquad (5.6)$$

Also, we may define a series of auxiliary variables

$$\left. \begin{array}{l} y_1 = y \\[2mm] \dfrac{dy_1}{dt} = y_2 = \dfrac{dy}{dt} \\[3mm] \dfrac{dy_2}{dt} = y_3 = \dfrac{d^2 y}{dt^2} \\[3mm] \vdots \\[2mm] \dfrac{dy_{n-1}}{dt} = y_n = \dfrac{d^{n-1} y}{dt^{n-1}} \end{array} \right\} \qquad (5.7)$$

and by substituting the equations (5.7) into (5.6), we convert equation (5.6) into a first-order differential equation in terms of the input $f(x)$ and the auxiliary variables:

$$b_n \frac{dy_n}{dt} = f(x) - b_{n-1} y_n - \ldots - b_1 y_2 - b_0 y_1 \qquad (5.8)$$

Dividing equation (5.8) throughout by b_n gives

$$\frac{dy_n}{dt} = \frac{f(x)}{b_n} - \frac{b_{n-1}}{b_n} y_n - \ldots - \frac{b_1}{b_n} y_2 - \frac{b_0}{b_n} y_1 \qquad (5.9)$$

Equations (5.7) and (5.9) constitute the decomposed set of equations that are equivalent to equation (5.6). So far, we have not specified what form the input term $f(x)$ will have. If we happen to be dealing with the decomposition of a transfer function as in equation (5.1), then we may work along roughly the same lines. Let us keep for the moment the Laplace notation of the transfer function, for being algebraic it is easier to manipulate than time-domain derivative terms. Let us begin by partitioning equation (5.1) to give

$$G(s) = \frac{Y(s)}{X(s)} = \frac{Y(s)}{U(s)} \cdot \frac{U(s)}{X(s)} \qquad (5.10)$$

where

$$\frac{U(s)}{X(s)} = \frac{1}{b_n s^n + b_{n-1} s^{n-1} + \ldots + b_1 s + b_0} \qquad (5.11)$$

and

$$\frac{Y(s)}{U(s)} = a_m s^m + a_{m-1} s^{m-1} + \ldots + a_1 s + a_0 \tag{5.12}$$

If we then clear fractions in equation (5.11) and re-arrange to solve for the highest power of s (corresponding to the highest derivative), we obtain

$$s^n U = \frac{1}{b_n} x - \frac{b_{n-1}}{b_n} s^{n-1} U - \ldots - \frac{b_1}{b_n} s U - \frac{b_0}{b_n} U \tag{5.13}$$

Let us now define the auxiliary variables

$$
\left.
\begin{aligned}
U_1 &= U \\
sU_1 &= U_2 = sU \\
&\vdots \\
sU_{m-1} &= U_m = s^{m-1} U \\
sU_m &= U_{m+1} = s^m U \\
&\vdots \\
sU_{n-1} &= U_n = s^{n-1} U
\end{aligned}
\right\} \tag{5.14}
$$

which when substituted into equation (5.13) give us

$$sU_n = \frac{1}{b_n} X - \frac{b_{n-1}}{b_n} U_n - \ldots - \frac{b_1}{b_n} U_2 - \frac{b_0}{b_n} U_1 \tag{5.15}$$

If we write equation (5.12) as

$$Y = a_m s^m U + a_{m-1} s^{m-1} U + \ldots + a_1 s U + a_0 U \tag{5.16}$$

then we may substitute into equation (5.16) the first m derivatives as defined in Laplace variable terms by the equations (5.14), to give

$$Y = a_m U_{m+1} + a_{m-1} U_m + \ldots + a_1 U_2 + a_0 U \tag{5.17}$$

If $m = n$ then all n derivatives of U are substituted.

The partition method is one of the most useful decomposition techniques for transfer functions, and is efficient in dealing with multivariable transfer functions with a denominator that is common to all the terms in the transfer function matrix.

There are other methods for the decomposition of transfer functions, each of which depends upon the definitions chosen for the set of auxiliary variables. Most commercially available simulation packages offer facilities for carrying out decomposition automatically, without the user having to know how it is done. It is probably sufficient for us, therefore, to view transfer functions as a special case in the decomposition of high-order differential equations.

Example

$$G(s) = \frac{Y(s)}{X(s)} = \frac{s^3 + 4s^2 + 2s + 8}{s^6 + 13s^5 + 52s^4 + 116s^3 + 176s^2 + 128s + 64} \tag{5.18}$$

If we partition equation (5.18) into

$$\frac{U(s)}{X(s)} = \frac{1}{s^6 + 13s^5 + 52s^4 + 116s^3 + 176s^2 + 128s + 64} \tag{5.19}$$

and

$$\frac{Y(s)}{U(s)} = s^3 + 4s^2 + 2s + 8 \tag{5.20}$$

then if we re-arrange equation (5.19) to solve for the highest power of s in U, we obtain

$$s^6 U = X - 13s^5 U - 52s^4 U - 116s^3 U - 176s^2 U - 128sU - 64U \tag{5.21}$$

If we then define the auxiliary equations

$$\left.\begin{array}{l} U_1 = U \\ sU_1 = U_2 = sU \\ sU_2 = U_3 = s^2 U \\ sU_3 = U_4 = s^3 U \\ sU_4 = U_5 = s^4 U \\ sU_5 = U_6 = s^5 U \end{array}\right\} \tag{5.22}$$

and write equation (5.20) as

$$Y = s^3 U + 4s^2 U + 2sU + 8U \tag{5.23}$$

then we may substitute equation (5.22) into equations (5.21) and (5.23) to give the complete set of equations

$$\left.\begin{array}{l} sU_6 = X - 13U_6 - 52U_5 - 116U_4 - 176U_3 - 128U_2 - 64U_1 \\ sU_5 = U_6 \\ sU_4 = U_5 \\ sU_3 = U_4 \\ sU_2 = U_3 \\ sU_1 = U_2 \\ Y = U_4 + 4U_3 + 2U_2 + 8U_1 \end{array}\right\} \tag{5.24}$$

Although there are six differential equations arising from the order of the transfer function denominator, the state variables U_1 to U_4 only are required to define Y.

5.3 USER INTERACTION

In our continuing critical examination of the use of digital computers for continuous-variable simulation, we need to cover user interaction with the model implementation, and how we may set about obtaining the highest degree of interaction possible under the circumstances.

When analog-minded engineers made the switch to digital computation for simulation purposes, they noticed a sharp drop in their interaction with their models. The sheer remoteness of the mainframe computers of the day, together with the unfriendly nature of data input and results output, gave very little room for good interaction in the simulation process. Engineers hitherto used to 'tweaking' their model parameters in rapid succession to effect improvements or try out different strategies, suddenly found that 'turn-around' times of a day or more from input of data to receipt of results were not uncommon. The greatest improvement came about when batch operation gave way to multiuser time-sharing computers with graphics visual display unit (VDU) screens. A simulation-ist could input data to a machine which gave the impression of being dedicated to his or her task, even though it was servicing the needs of many other users. It was only when heavily loaded that time-sharing mainframes gave poor response times. In the digital world, the latest (and best) hardware platforms for working are the so-called personal computer (often called a PC) and the graphics workstation. Developments are such that the computing power of these machines continues to increase with every new model, and it is difficult to assess what the limits might be for this technology.

User interaction is obviously at its best on a dedicated machine, and therefore a single-user personal computer or workstation that has the right specification is going to give the best possible service. Machines of this type that are part of a network and depend on a file server will suffer the disadvantages of a time-sharing mainframe if the system is heavily loaded. The best interaction will be obtained under the following conditions:

- The simulation language coding statements are easy to edit. It is preferable in many instances if an on-line text editor is available from within a package, for this saves the overhead of exit from the package, editing and then re-entry for a new run.
- The simulation language being used 'sits on top of' (i.e. interfaces with) a fast-acting high-level language compiler. Most simulation packages translate into FORTRAN, so rapid compilation is essential for fast turnround in model development.
- The simulation language translator is able to effect speedy production of the FORTRAN subroutines equivalent to the language coding statements that describe the model.
- Some simulation packages have interpretive facilities. This means that each coding statement or instruction may be translated individually and

immediately to a machine-executable form, thus avoiding the need for a compiler. This has advantages for greater machine interaction, but if the execution of the simulation is wholly interpretive, speed of execution may suffer.

- The facilities for data input prior to a simulation run should be as user-friendly as possible, that is with error trapping and on-line help being available.

- Facilities should be available for obtaining graphical output preferably at any stage of the simulation and not just upon completion of a run. Run-time graphical output can enable early diagnosis of incorrect run conditions, and allow timely termination of an abortive run.

5.4 COMPREHENSIVE USE OF GRAPHICS

Analog computer operation depends upon the production of graphical output (either on a display oscilloscope or on an $X–Y$ plotter) for communication with the user. The spin-off from this form of output is its high information content and the consequent benefit to the simulationist, particularly during model development. The speedy production of adequate graphics needs to be part and parcel of a digital simulation package, if it is to be of greatest advantage to the user of such a facility. There are various aspects of graphics that need to be addressed separately, in order to get the best out of the facility.

5.4.1 Run-time Graphics

The provision of graphical output at run time is a feature that is somewhat in dispute. Some authorities do not consider it necessary to be able to view even some of the key transients, and there are simulation packages that do not provide this facility.

There are advantages in having run-time plot facilities. The greatest of these lies in being able to detect, in the early stages of a run, the presence of any malfunction of the simulation. A simulation run with a large model can take a long time to run its full course, and early detection of incorrect operation can enable the run to be aborted without wasting computing time to no avail.

Run-time plotting can be very useful too in the kind of highly interactive work that requires a simulationist to experiment with a succession of parameter changes, one after the other. Observation of the effects of the changes at run time can enable a simulationist to make decisions for further runs, without having to indulge in post-run plotting for the results. In certain circumstances, this may be carried further by having a run interrupt facility which allows parameters to be altered during a run, before resuming to see the consequences of the changes.

If facilities are available for obtaining run-time graphics, you should not try to plot everything in sight. It is normally sufficient for diagnostic purposes to choose a few key variables which are likely to exhibit behaviour highly indicative of the state of the model. In calling for run-time graphics plots, you will experience a throwback to the days of analog computing! You will need to estimate maximum and minimum values for the scales on your plots, since these will not be known before the run has taken place.

5.4.2 Post-run Graphics

After the run has finished, all the run data is available and post-run graphics may be obtained for all the chosen transients. For efficiency, it is usual for the run-time executive of the simulation package to be notified as to which variables are to be monitored for post-run output. During the course of development work on a model, it is common to overspecify the list in an attempt to form proper judgements about behaviour, but for a model that is working to the satisfaction of the user, the selection will often be somewhat narrower.

In post-run graphics output, it is very useful to be able to indulge in what may be called *graphics editing*. This consists of 'playing around' with plots in order to get the most useful information out of them. Scale alteration, both for time and for dependent variables; plotting several transients on individual or identical scales; plotting one or more dependent variables against another dependent variable – these are all examples of the way that graphics may be manipulated. For this reason, post-run graphics should be an interactive process, allowing the user to issue commands for the specification of plots.

5.4.3 Special Forms of Graphics

Up to now, the graphics output has been in the form of transients depicting model-variable behaviour. There are occasions where graphics of a different kind are required, outside the scope of transient plotting. Examples of this type are given below:

- The results of a parameter survey over several runs, where a specified result for each run is plotted against the corresponding parameter value. A simple example of this is a simulation of a test firing facility, where a series of runs with different firing angles gives rise to a plot of missile range against angle. This can then be used for preliminary calibration purposes.
- In simulation involving stochastic processes, the transient data obtained may be subjected to spectral analysis, and functions such as power spectral density or various correlations may be plotted.

5.5 SIMULATION STRUCTURE FOR DIGITAL COMPUTATION

Simulation structure is fundamentally the same as the general outline given in Section 2.4. Any differences from analog computation lie in the method of implementation at each stage, rather than in the overall philosophy behind the concept of structured simulation. Figure 5.1 shows a diagram of overall simulation structure, which draws together the different procedures within the three stages, for all three types of computer and for both continuous and discrete simulation. Although, as shown, some of the procedures apply only to analog and hybrid computers, it is worth considering whether they have any value for digital computer implementation.

5.5.1 Model Set-up and Initialization

Setting up a digital implementation of a simulation may be considered to be two separate tasks, as indicated in Figure 5.1. The problem set-up consists of coding the model equations in the language medium being used, and perhaps specifying other model elements such as fixed constants (which are not expected to change from run to run). The parameter set-up requires specification of those elements which are expected to change from run to run. This may include the initial values for the differential equation (state) variables. For greatest efficiency of operation, parameter set-up and alteration should be under the control of the run-time executive for the package being used. By this means, a series of runs with a particular model will be executed in succession without the need for recompilation between run changes.

5.5.2 Getting the Model to Run

Let us assume that we have set up a simulation model for running with a simulation package. The model description coding statements have been entered, all constants, parameters and initial conditions have been specified, and the model is now ready to run. We find all too often that when we come to try to get the simulation to run (i.e. execute the dynamic phase), it soon fails on a system run-time error. What can we do to check our model implementation to try to obviate such occurrences?

Once again, it may be useful to take a leaf out of the 'book' of analog computation. Two procedures that are more or less essential to the successful set-up of an analog (and the analog part of a hybrid) computer model are the *static* and *dynamic checks*. To recapitulate, the static check is the procedure of monitoring the outputs of all the amplifiers 'patched' together to form the model implementation. This is done in initial conditions mode, or by the application of arbitrary values to the initial conditions inputs of all the integrators. The dynamic

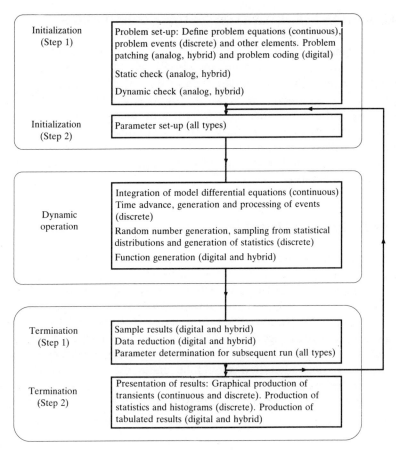

Figure 5.1   Overall simulation structure.

check is a testing sequence to ensure that the analog integrators are all functioning as expected.

Let us now examine the digital model implementation to see if there is any comparable procedure that may help us check the model operation. After the initialization phase is complete, and the model is ready for the dynamic phase to begin, it is very useful to carry out just one pass through the complete set of model equations, at simulation time zero, and with the initial conditions applied. To be most effective, the simulation package should be able to execute in 'debug mode' with full access to the debugging utility that goes with the FORTRAN compiler being used. Alternatively, the simulation package itself should be able to operate in this mode, with the ability to step through the execution of statements, one by one, and with access to the values of all variables. Checking the model equations in this way will often point to incorrect results being obtained from certain equations. The coding can be checked and the necessary model alterations made. Another advantage that accrues from this form of examination, which verges on a dynamic check, is that unexpectedly high or low values for system derivatives may be revealed.

Taking this checking process a stage further leads us to the digital counterpart of the dynamic check. Assuming that the single pass through the model equations revealed no glaring errors, the next procedure to adopt is to try a very short simulation run to test dynamic operation. If there are any peculiarities of dynamic behaviour, they will very often appear right at the beginning of the run. If all is well, then the time has come to be adventurous and try for a full-length run.

5.5.3 Checking and Analyzing the Results

The first full-length run that should be attempted is one in which the model is allowed to 'float'. In this mode of operation, no input stimuli (forcing functions) are applied to the model, so that certain inherent properties of the model itself may be investigated:

- Whether or not the initial state of the model is a steady state.
- Whether the model behaviour diverges from, or converges towards, a steady state.
- If there is a slow 'drift' of the model away from the initial state (frequently the case for non-linear models).

In this last case, the steady state of the model may not be known, and a special technique known as *trimming* is used to achieve steady-state behaviour. This constitutes the judicious application of (usually small) inputs to certain points in the model to counteract the drift and bring about a steady state. If the unforced behaviour of the system model appears satisfactory, it is then usually all right to proceed with model testing, forcing functions being applied as required.

When it comes to the transient testing of models, it is difficult to generalize with regard to test methods. It usually comes down to taking a long hard look at the model transient responses obtained from the application of forcing functions, and deciding upon the basis of one's own skill and judgement whether the results are acceptable or not. This is best done by making sure that all the relevant variables are output in graphical and (for detailed checking) tabulated printout form. In large simulation models, it is usually easier to test subsections of the model separately before linking them together to form the whole. The examples later in the chapter will hopefully serve to show the application of some of the above principles.

5.6 IMPLEMENTATION PROBLEMS NOT FOUND WITH ANALOG COMPUTATION

Although analog computing is rarely used in present times, there are lessons to be learned from the manner in which models are organized for analog computation, particularly when it comes to overcoming some problems that occur in digital simulation that are not found in analog computation.

5.6.1 Digital Computation Coding Sequence

As we saw in Chapter 4, an analog computer is a parallel device, in that all variables are computed simultaneously. This truly represents the state of affairs in the continuous system being modelled. In performing a digital simulation of a continuous process, we have to contend with the fact that a digital computer is a serial computing device in which operations are performed in sequence. In order to meet the same requirements for its closed loops as the analog computer, we need to ensure that a variable is not used in a computer language statement unless it has been defined beforehand. This involves sorting the statements that define the model structure into the correct sequence for computation. If this statement ordering can be carried out by the software of a digital simulation package, then the sequential digital computation appears to the user to be a parallel operation, because statement ordering is no longer of any consequence.

There are, however, some differences from analog computation. The first is that, although the loops are broken at time $t = 0$ for the application of initial conditions in the same way as on an analog computer, the loops remain broken in the same places throughout the simulation. This is because the derivative of a variable (corresponding to the sum of inputs to an analog integrator) is defined in the long run as a function of integrator output values given by the initial conditions at time $t = 0$, or by the integration algorithm at time $t > 0$.

For example, let us consider the single differential equation

$$y' = f(y) \tag{5.25}$$

At time $t = 0$ (from equation (5.25))

$$y'_0 = f(y_0) \tag{5.26}$$

If we consider that the integration algorithm uses the value y_0, the value of time t_0, the derivative at time $t = 0$ and the size of the time step, h, away from the initial condition, to compute a new value of y at the end of the step we may write

$$y_1 = g[y_0, t_0, y'_0, h] \tag{5.27}$$

for the first step of the solution. This procedure is repeated for the second and subsequent steps until at time $t = nh$ (where h is still the step length)

$$y'_n = f(y_n) \tag{5.28}$$

$$\therefore \quad y_{n+1} = g[y_n, t, y'_n, h] \tag{5.29}$$

Thus in a digital computer simulation, the integration algorithm is relied upon to close the loops around the integrators under all conditions.

The second difference between digital and analog computation lies in the treatment of algebraic loops. In a continuous system simulation, variables are often defined by algebraic rather than by differential equations, and it is where calculation loops exist between sets of such variables that trouble can occur. In analog computation, algebraic loops are permissible because of the truly parallel nature of an analog computer. In digital simulation, however, algebraic loops are not permitted unless special devices are used to cope with the situation.

Algebraic loops are of two kinds, those involving algebraic variables only, and those involving derivatives as well as algebraic variables. Examples of algebraic loops are

- *Algebraic variables*

$$y = a^x + b \tag{5.30}$$

$$z = cy + d \tag{5.31}$$

$$x = \sin(z) + y \tag{5.32}$$

This loop may be dealt with by incorporating these equations into an iterative sequence, with an initial value given for one of the variables. This value is changed at each iteration until it converges to the point at which all equations in the loop are satisfied.

- *Derivatives*

$$\frac{dx}{dt} = x \, \frac{dy}{dt} \tag{5.33}$$

$$\frac{dy}{dt} = y + \frac{dx}{dt} \tag{5.34}$$

The loop in this case may be removed by substituting for dx/dt in equation (5.34), to give

$$\frac{dy}{dt} = \frac{y}{1 - x} \tag{5.35}$$

which, followed by equation (5.33), gives the correct computation sequence.

Derivatives are, in effect, algebraic variables, and could have appeared in either of the above examples. Sets of linear equations can always be substituted to remove algebraic loops. Non-linear equations need the use of special devices to deal with the situation.

A simulation package should warn of the presence of algebraic loops and offer the means for coping with them. ACSL will sort model coding statements into the correct computational sequence (unless specifically prevented from doing so by the use of *procedural* blocks), and has devices for dealing with algebraic loops, not least of which is the *implicit solution of DAEs*, covered in the next chapter.

5.6.2 Numerical Integration Algorithms

The analog computer has but one means of carrying out the operation of integration, namely by the charging and discharging of capacitors in feedback around the integrators. This integration will proceed at the same rate, no matter how large the simulation problem in hand, and irrespective of the nature and speed of the solution transients. Digital simulation, on the other hand, is very sensitive to the nature and size of the simulation model, and many different integration algorithms have been developed to cope with the wide variety of situations that may arise in simulation. The subject of the numerical analysis of integration is covered more fully in the next chapter.

5.6.3 Discontinuities

In analog computation, when a switching transient takes place, the result is a discontinuity in the solution profiles. This has no effect whatsoever upon the speed of solution. Digital integration, however, is very sensitive to the presence of discontinuities, unless certain types of algorithms are used which are capable of negotiating discontinuities with minimum disruption. The subject of discontinuities is dealt with fully in Chapter 7.

5.6.4 'Stiffness' in Sets of Differential Equations

The solution to a differential equation may be such that the solution variable changes very slowly or very rapidly with respect to time. A parameter known as the *time constant* characterizes the rate of change of the solution variable in response to inputs to the equation. If a problem contains differential equations having a wide range of time constants, then it will more than likely have an adverse effect upon numerical integration speed. The analog computer does not suffer unduly from this phenomenon, known as the property of *stiffness*. Section 6.3 in the next chapter deals with methods for overcoming the effects of stiff equations on the speed of solution.

5.7 EXAMPLE: A SIMPLE CONTINUOUS FEEDBACK SYSTEM

5.7.1 The Model

Figure 5.2(a) shows a block diagram for a simple feedback system. For those readers not very familiar with control engineering terminology, the system is unity

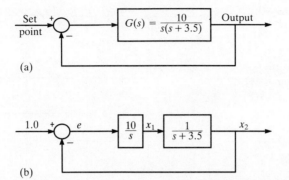

Figure 5.2   Block diagram of a simple feedback system, $a = 3.5$.

feedback (i.e. the output is fed directly back to a point of comparison with what is known as the *set point*). The set point may be fixed, or may be a time-varying quantity that represents the required pattern of behaviour of the system over a period of time.

Any difference between the required and actual pattern of behaviour (i.e. between the set point and the system output) actuates the process described by the *transfer function*

$$G(s) = \frac{K}{s(s + a)} \quad \text{where } K = 10 \text{ and } a = 3.5 \tag{5.36}$$

This is a *second-order* system where $G(s)$ is the Laplace transform representation of a pair of first-order ordinary differential equations that describe the dynamic behaviour of the system output in response to an input stimulus. The input stimulus, as we have seen, is provided by the difference between the required and actual patterns of behaviour, and ideally should serve to reduce that difference to a minimum.

For the purposes of simulation, we need to decompose the second-order process into two first-order ones for solution by means of a general-purpose simulation package. Figure 5.2(b) shows the block diagram with the transfer function split into two blocks, each of which represents a differential equation of the system. The set point is shown as having a value of 1.0. The equations may be written as

$$e = 1.0 - x_2 \tag{5.37}$$

$$\frac{dx_1}{dt} = 10e \tag{5.38}$$

$$\frac{dx_2}{dt} = x_1 - 3.5x_2 \tag{5.39}$$

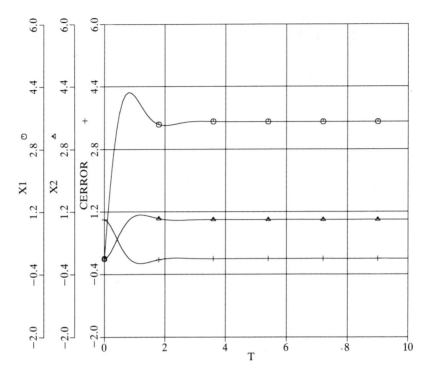

Figure 5.3   ACSL run-time plot of responses to a unit step, *a* = 3.5.

If the initial values of x_1 and x_2 are zero, then the use of a set point value of 1.0 is tantamount to the application of a unit step forcing function to the system. The system response will be a *unit step response*, used very commonly by control engineers for testing control systems. Intuitively, we already know that the system is going to be stable, and that 'floating' the system will produce no change whatsoever in the system behaviour. In this instance it is quite appropriate to institute the set point value from time zero, thereby introducing the unit step change at that point.

5.7.2 Simulation Results

Figure 5.3 shows a run-time plot for a simulation of this model behaviour, using the ACSL simulation package with a file called CONTIN.CSL. What this tells us is that the system is stable within the timescale of interest, in that its response to a unit step input is a stable transition to a steady state with output value reaching 1.0, exactly that required by the set point. Figure 5.4 shows a set of post-run transients obtained for the system responses. We should note that x_1 reaches a

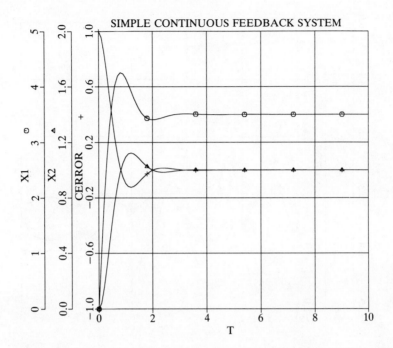

Figure 5.4   ACSL post-run plot of unit step responses, *a* = 3.5.

steady-state final value of 3.5, as x_2 reaches 1.0. This is correct, and in line with the value of *a* used in the model equations.

In keeping with the urge to experiment with our model, we may alter the value of *a* to see its effect. Figure 5.5 shows the run-time transient responses obtained from an ACSL run with *a* set to 0.5. The responses are now clearly very oscillatory, although the system is still stable and would settle to a steady state, given enough time. If we obtain as a post-run plot the response of x_2 plotted against x_1 instead of time, we obtain what is called a *phase plane plot* (Figure 5.6). It may be easier by looking towards the centre of the spiral to estimate the final values for the two variables. As x_2 approaches a value of 1.0, x_1 moves towards a value of 0.5, again in keeping with our new value of *a* in equation (5.25). A model statement listing for ACSL is given in Appendix A.1.

An equivalent model was written for implementation with the MATSIM routines. Details are given in Appendix A.2, together with listings of the M-files CONTIN.M (for set-up and initialization) and CONTB.M (the model equations). For comparison purposes, Figure 5.7 shows the run-time plots, and Figure 5.8 some post-run plots, obtained from the MATLAB implementation. The results from both model implementations are identical.

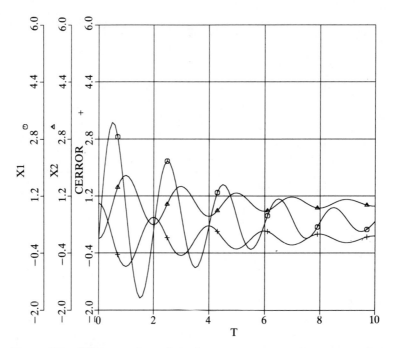

Figure 5.5   ACSL run-time plot of responses to a unit step, *a* = 0.5.

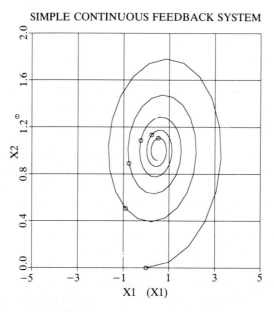

Figure 5.6   ACSL phase plane plot of responses, *a* = 0.5.

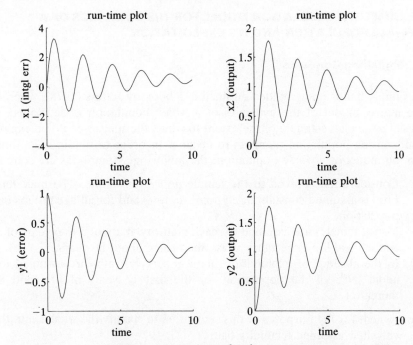

Figure 5.7   MATSIM run-time plots of unit step responses, *a* = 0.5.

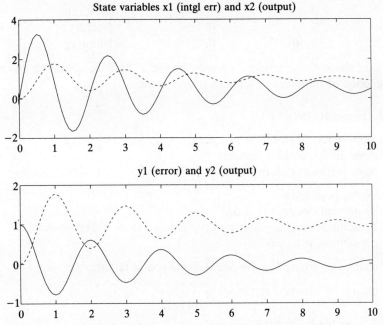

Figure 5.8   MATSIM post-run plots of unit step responses, *a* = 0.5.

5.8 EXAMPLE: A SIMULATION MODEL FOR THE DYNAMICS OF A WHALE POPULATION AND ITS EXPLOITATION

5.8.1 Modelling Concepts

This example builds upon an idea outlined in a book by Jeffers (1978), whereby a simple matrix model of the dynamics of a whale population is developed, and then used to give an estimate of the extent to which the species could be exploited without causing population numbers to decline. In order to obtain a continuous simulation model of a whale population, the following assumptions are stated:

- Consideration is given to the female population alone. The male/female ratio is assumed constant at all points in time and for all age groups in the population.
- Young females are assumed to reach maturity at about 5 to 6 years of age, and natural life expectancy is assumed to be about 50 years.
- In the absence of exploitation, natural survival rates are assumed to be about 89% of the population for the first 12 years of life, and 82% thereafter.

It is convenient for the purposes of this simulation to classify the whales into three groups with their assumed fecundity figures:

Group	Age	Average fecundity	
1	0–4	0	(Young)
2	5–12	0.205	(Adult)
3	13–50	0.225	(Elderly)

For each of these three groups, we may derive a 'balance sheet' to describe the rate of change of numbers of females in the group.

Group 1. Young females (Y)

Input to population from births per year
  (a) from births to adult females $\qquad$ $0.205A$
  (b) from births to elderly females $\qquad$ $0.225E$

Removal from population
  (a) by natural deaths per year $\qquad$ $0.11Y$
  (b) by maturation into adult group (assumed to be 1/4 of young
      population per year) $\qquad$ $0.25Y$

Rate of change of young female population (young females per year) $\qquad$ dY/dt

Rate of change = inputs − removals

$$\therefore \frac{dY}{dt} = 0.205A + 0.225E - 0.36Y \qquad (5.40)$$

Group 2. Adult females (A)

Input to population from maturation of young
(1/4 of young population per year) $0.25Y$

Removal from population
  (a) by natural deaths per year $0.11A$
  (b) by maturation into next age group
     (1/8 of adult population per year) $0.125A$

Rate of change of adult population
(adult females per year) dA/dt

Rate of change = inputs – removals

$$\therefore \; \frac{dA}{dt} = 0.25Y - 0.235A \tag{5.41}$$

Group 3. Elderly females (E)

Input to population from ageing of adults
(1/8 of adult population per year) $0.125A$

Removal of population by natural deaths per year $0.18E$

Rate of change of elderly population
(elderly females per year) dE/dt

Rate of change = inputs – removals

$$\therefore \; \frac{dE}{dt} = 0.125A - 0.18E \tag{5.42}$$

We now have three differential equations which describe a model of the whale population dynamics in the absence of exploitation:

$$\frac{dY}{dt} = 0.205A + 0.225E - 0.36Y$$

$$\frac{dA}{dt} = 0.25Y - 0.235A$$

$$\frac{dE}{dt} = 0.125A - 0.18E$$

If we now consider exploitation of the population, we may define an exploitation factor F as the fraction of whales removed each year by whaling. If we assume

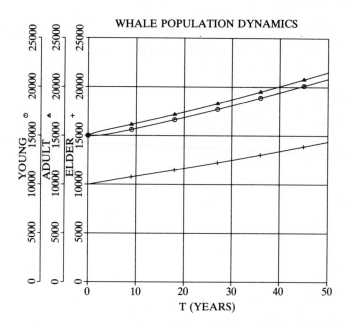

Figure 5.9   ACSL run of whale population model: transient responses of population without exploitation.

that this fraction applies equally to each of the three groups, and to males and females alike, we then obtain the modified equations:

$$\frac{dY}{dt} = 0.205A + 0.225E - 0.36Y - F \cdot Y \tag{5.43}$$

$$\frac{dA}{dt} = 0.25Y - 0.235A - F \cdot A \tag{5.44}$$

$$\frac{dE}{dt} = 0.125A - 0.18E - F \cdot E \tag{5.45}$$

The equations may now be used to study the consequences of exploitation of the whale population.

5.8.2 Running the Model

A program (WHALE.CSL) has been written for the ACSL simulation package, the full listing for which is given in Appendix A.1. Figure 5.9 shows a plot of the

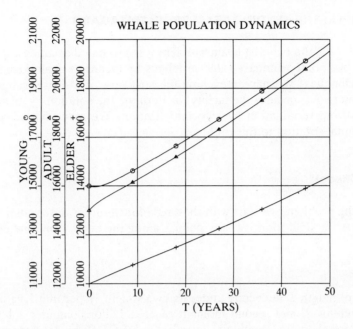

Figure 5.10   ACSL run of whale population model: no predefined scaling.

transient response for the three age groups over a period of 50 years, when it is assumed that no exploitation of the species takes place. In the absence of any predation by humans, the model output shows a steady growth of the population, which, on the basis of model predictions, would double in about 120 years, so there is scope for some exploitation without causing the species to decline. Figure 5.10 shows a plot of the same output, this time without any predefined scaling, so that the transients take up a greater range on the *Y*-axis scale in each case.

5.8.3 Exercises

(a)   The initial conditions for the simulation can be seen to have been chosen as $A = 15000 = Y$, $E = 10000$. Is there a basis for this choice? Try different sets of initial values and observe the results.

(b)   Re-run the simulation with non-zero values for the exploitation factor F (called FACTOR in the program listing). Find a value that just allows positive growth of the population to take place. Plots as shown in Figure 5.10 are useful, owing to the greater sensitivity of scaling available. Note the constancy of the ratios that the three age groups have with respect to each other. This is to be expected from a linear model like this one.

5.9 EXAMPLE: THE BENDING BEAM PROBLEM (AGAIN)

In Section 4.3.1.2 the bending beam problem was presented as an example of the solution of two-point boundary-value problems by means of analog computation iterative techniques. We have seen how the structure of a digital simulation as shown in Figure 5.1 includes the facility for iterating the solution by means of the return loop from termination back to initialization. We may take advantage of this for a digital solution to this problem.

5.9.1 The Model

Re-stating the problem, we start with the single fourth-order differential equation that describes the deflection y at any point x along the beam from one end:

$$EI \frac{d^4 y}{dx^4} = -W \tag{4.41}$$

for a beam of length L deflecting under its own weight W per unit length, having Young's modulus E and second moment of area I. For computer solution, we need to re-formulate the equation as four first-order ODEs by decomposition into four state variables (just as we did in Chapter 4):

$$\frac{dy_4}{dx} = -\frac{W}{EI} \tag{4.43}$$

$$\frac{dy_3}{dx} = y_4 \tag{4.44}$$

$$\frac{dy_2}{dx} = y_3 \tag{4.45}$$

$$\frac{dy_1}{dx} = y_2 \tag{4.46}$$

We assume, once again, that the beam is embedded at both ends in supporting walls, so we may define boundary conditions

$$y_1 \ (y) = 0 \text{ at } x = 0 \text{ and at } x = L \tag{4.51}$$

$$y_2 \ (\dot{y}) = 0 \text{ at } x = 0 \text{ and at } x = L \tag{4.52}$$

For convenience, these boundary conditions will be re-stated as

$$y_1(0) = y_1(L) = 0.0 \tag{5.46}$$

$$y_2(0) = y_2(L) = 0.0 \tag{5.47}$$

$y_3(0)$ and $y_4(0)$ are unknown and must be determined so as to satisfy the boundary conditions at $x = L$ for y_1 and y_2. This is done by using a sensitivity matrix approach as follows:

- One run is performed with all initial values set to zero. The final values for y_1 and y_2 are saved as $y_{10}(L)$ and $y_{20}(L)$. The second subscript denotes that no perturbations were made to the initial conditions for this run. The initial values $y_{30}(0)$ and $y_{40}(0)$ are also saved (unperturbed).
- The initial value $y_3(0)$ is perturbed and another run carried out to produce the final values $y_{13}(L)$ and $y_{23}(L)$ which are also saved. The second subscript indicates that these final values are the result of perturbing $y_3(0)$.
- The initial value $y_3(0)$ is reset to its unperturbed state, and $y_4(0)$ is perturbed. The final values $y_{14}(L)$ and $y_{24}(L)$ are obtained from a third simulation run, and saved.
- A sensitivity matrix

$$A = \begin{bmatrix} a_{13} & a_{14} \\ a_{23} & a_{24} \end{bmatrix} \tag{5.48}$$

is computed from the relationships

$$a_{ij} = \frac{\partial y_i(L)}{\partial y_j(0)} \tag{5.49}$$

in terms of partial derivatives denoting change in final value $y_i(L)$ per unit change in initial value $y_j(0)$.
- The required change to the initial conditions vector is given by

$$\begin{bmatrix} \Delta y_3(0) \\ \Delta y_4(0) \end{bmatrix} = A^{-1} \begin{bmatrix} \delta y_1(L) \\ \delta y_2(L) \end{bmatrix}$$

$$= A^{-1} \begin{bmatrix} -y_{10}(L) \\ -y_{20}(L) \end{bmatrix} \tag{5.50}$$

since the required final values vector is

$$\begin{bmatrix} 0 \\ 0 \end{bmatrix}$$

This yields a new initial-values vector needed for (hopefully) the final run, if the solution converges to the correct final values. If it does not (likely with non-linear problems), then the sequence of simulation runs is repeated from this new starting state with smaller perturbations, on the assumption that the initial values are near to those required. For this linear problem, the sensitivity matrix approach should yield the exact solution after just three iterations (i.e. four runs in all).

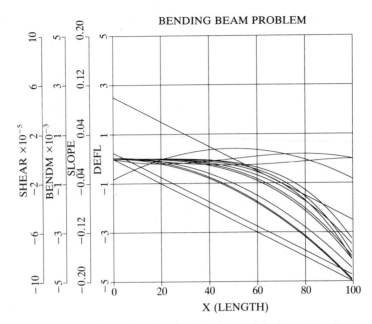

Figure 5.11   ACSL run of bending beam model: all iteration transients shown on one set of axes for run-time plot.

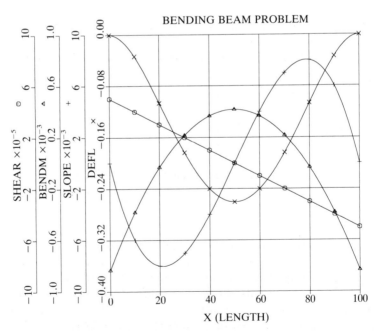

Figure 5.12   ACSL post-run plot for bending beam problem, showing final solution after four iterations.

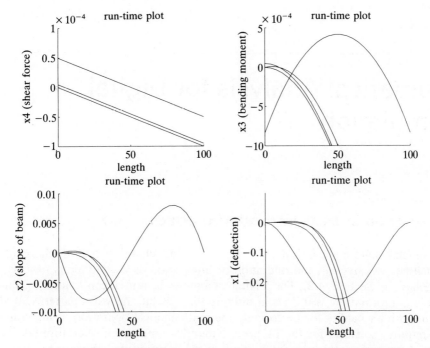

Figure 5.13   MATSIM run-time plot for bending beam problem.

5.9.2 Simulation Results

A listing for an ACSL implementation of this problem is given in Appendix A.1 (files BEAM.CSL and BEAM.CMD). Figure 5.11 shows, on one set of axes, all the iterations that took place to produce the final solution. This is shown in Figure 5.11, but is re-presented in Figure 5.12 as a result of running the problem again. This time it only runs once because the correct initial conditions have already been found from the previous iterations. This is a trivial example of an iterative solution to a problem being handled by digital methods, but serves to illustrate the manner in which convergence may be accelerated by techniques much more sophisticated than those available with analog computation. Figure 5.13 shows a run-time plot obtained from the MATSIM with the beam model BEAMIN.M and BEAMB.M (listed in Appendix A.2). This shows better the iteration stages that have to take place to produce the final result.

5.10 REFERENCE

JEFFERS, J N R (1978) *An introduction to systems analysis: with ecological applications*, London: Edward Arnold.

6

Numerical Analysis for Digital Simulation

6.1 WHAT DO WE MEAN BY NUMERICAL INTEGRATION?

The analog computer is able, by virtue of the charging and discharging of capacitors, to carry out the operation of integration as a continuous process, as described in Section 4.2. The digital computer is not a continuous machine, handling information as it does in discrete form as bit and word patterns. It also has no in-built facility for performing the mathematical operation of integration, and requires software for the purpose. Numerical integration therefore takes the form of processing, through the medium of the appropriate software, to carry out the operation of integration in a quasi-continuous way.

6.1.1 Taylor Series as the Starting Point

In continuous-variable simulation, the Taylor series is a mathematical expression for computing the next value in a time series of values that constitute the solution to a differential equation of any order. Instead of being able to advance the solution in a continuous way, as in analog computation, we need to advance the solution incrementally with respect to time. If the time increments are small enough, the solution looks continuous, especially if the values obtained are joined together by small vectors as piecewise linear plots. For any order of differential equation, the solution technique is as follows:

- Starting from a given value of the solution (which will be the initial value at time zero), we wish to advance the solution to a new value after an increment of time has elapsed.
- For a differential equation of any order, the first derivative (if present) represents the rate of change of the variable. Its contribution to the new value at the end of the time step will be given by multiplying this rate of change by the time increment itself.
- The second derivative (if present) represents a rate of change of the first derivative. Its contribution is given by multiplying the time interval by the

difference between first-derivative contributions at the beginning and end of the time step.

- This procedure is continued adding in the contributions made by each of the derivatives present, to obtain the value at the end of the time step, correct according to the order of the differential equation being solved.

This procedure may be stated mathematically in the form of the *Taylor series*

$$x(t + h) = x(t) + hx'(t) + \frac{h^2}{2!}x''(t) + \ldots + \frac{h^n}{n!}x^{(n)}(t) + \ldots \tag{6.1}$$

where t is time, h is a time increment, x is the solution variable of the differential equation being solved, x' is its first derivative, x'' is its second derivative, $x^{(n)}$ is the nth derivative. Also $x(t)$ is the value of x at the beginning of the time increment, while $x(t + h)$ is the value at the end, given as $x(t)$ plus all the contributions from the derivatives of x.

Most numerical integration methods are derived from the Taylor series. In common with analog computation, however, all our differential equations to be solved will be first order and therefore of the form

$$x'(t) = f(t, x(t)) \tag{6.2}$$

with an initial value

$$x(0) = x_0 \tag{6.3}$$

The problem that arises when equation (6.1) is applied to numerical integration (i.e. time advance from t to $t + h$) is that of calculating the values of second- and subsequent-order derivatives, when all that we have is a first derivative for each differential equation present in our model. We have seen how high-order differential equations may be decomposed into sets of first-order ordinary differential equations. What we do not know, however, is the form of any high-order differential equation that may replace a set of first-order equations, unless we have a linear model. Hence, from this point of view as well, the higher derivatives are not usually available to us. Our task in using the Taylor series as the basis of numerical integration lies in obtaining suitable finite-difference representations for the second- and higher-order derivatives, where the only information we have is the values of the solution variables, together with their first derivatives. A reasonably simple classification of numerical integration methods, therefore, is based upon the type of approximations adopted for these higher-order derivatives.

Let us adopt a slightly different notation whereby

$$x_n = x(t) \tag{6.4}$$

$$x_{n+1} = x(t + h) \tag{6.5}$$

$$x'(t) = f(t, x(t)) = f(t, x_n) \tag{6.6}$$

We may therefore express equation (6.1) as

$$x_{n+1} = x_n + hf(t,x_n) + \frac{h^2}{2!} f'(t,x_n) + \ldots + \frac{h^n}{n!} f^{(n-1)}(t,x_n) + \ldots \qquad (6.7)$$

For the purposes of this discussion, classification is based upon two sets of choices that have to be made:

- Single-step or multiple step methods
- Use of explicit or implicit formulae.

It is probably simpler to deal with these under different headings for single- and multiple step methods, and to examine the differences for explicit and implicit versions of the formulae, within these two headings.

6.1.2 Single-step Integration Methods

Commonly known collectively as *Runge–Kutta* methods, this class of numerical integration methods depends upon 'matching' or approximating the derivative terms in the Taylor series by a number of first-derivative evaluations at various points *within* the current step h, followed by a weighted averaging of these evaluations. This is done either according to the *explicit* formula

$$\left.\begin{array}{l} k_1 = hf(t,x_n) \\[2mm] k_r = hf\left(t + a_r h, x + \displaystyle\sum_{s=1}^{r-1} b_{rs}k_s\right) \quad r = 2,3,4,\ldots,m \\[4mm] x_{n+1} = x_n + \displaystyle\sum_{r=1}^{m} c_r k_r \end{array}\right\} \qquad (6.8)$$

or to the *implicit* formula

$$\left.\begin{array}{l} k_r = hf\left(t + a_r h, x + \displaystyle\sum_{s=1}^{m} b_{rs}k_s\right) \quad r = 1,2,3,4,\ldots,m \\[4mm] x_{n+1} = x_n + \displaystyle\sum_{r=1}^{m} c_r k_r \end{array}\right\} \qquad (6.9)$$

The *order* of the method m reflects the point at which the Taylor series is truncated before any matching is attempted.

Each k-value is an estimate of the increment to be added to x_n to produce the value x_{n+1} at the end of the step. In the explicit Runge–Kutta algorithm, each succeeding k-value is based upon the results obtained in the preceding one. k_1 is based upon the value of the first derivative at the beginning of the step, while subsequent k-values are computed at points within the step, at fractions a_r of the step of size h, and as functions of x that are themselves based on other estimates of k. One pass through all the formulae will produce all the k-values.

In the implicit algorithm, the k-values are obtained from what are called *closed formulae* in which the solution variable appears on both sides of the equation. In order to solve a closed formula, where it is not possible to group terms in the solution variable onto one side of the equation, it is necessary to adopt some *iterative process* such as *Newton–Raphson* to obtain a solution. In the fully implicit Runge–Kutta method, it is strictly speaking necessary to iterate to convergence for all the k-values at the same time. This will introduce a considerable computational overhead at each integration step, so a compromise that is often adopted is to use a *semi-implicit* Runge–Kutta formula, where the closed formula approach is confined to a subset (often only one) of all the k-values (Rosenbrock, 1963, Caillaud and Padmanabhan, 1971).

Truncation of the Taylor series, which defines the order of the method, leads to a *truncation error* in the application of a Runge–Kutta method over an integration step. Let us consider the two simplest forms of Runge–Kutta algorithm, the first- and second-order methods, and examine their operation through the medium of a geometrical construction. The first-order formula (also known as the *forward Euler formula*) may be written as

$$\left. \begin{array}{l} k_1 = hf(t, x_n) \\ x_{n+1} = x_n + k_1 \end{array} \right\} \quad (\text{error} = O(h^2)) \tag{6.10}$$

In Figure 6.1, for a differential equation $\dot{x} = f(t, x)$, the slope is evaluated in terms of conditions at the beginning of the step, and applied throughout the whole step of size h to give k_1 at the end of the step. This is the increment that is added to x_n to give a first-order estimate for x_{n+1}.

The first-order Runge–Kutta formula represents the Taylor series with all terms after the second removed. These are terms that contain h^n ($n \geqslant 2$), and hence the shorthand form of the truncation error is expressed as $O(h^2)$ (an error of the order of h^2).

Let us now apply a second-order Runge–Kutta formula, and examine its geometry. The formula is given as

$$\left. \begin{array}{l} k_1 = hf(t, x_n) \\ k_2 = hf(t + \alpha h, x_n + \alpha k_1) \\ x_{n+1} = x_n + c_1 k_1 + c_2 k_2 \end{array} \right\} \quad (\text{error} = O(h^3)) \tag{6.11}$$

As with the first-order method, k_1 is evaluated, but this time it is one of two estimates for k. The second estimate is determined as follows (see Figure 6.1):

- After determining k_1, a value α is chosen ($0 < \alpha < 1$), from which a point in time within the current step is calculated as $t + \alpha h$.
- By similar triangles, the corresponding value of x on the k_1 slope is given as $x_n + \alpha k_1$.

- A new slope is calculated (as the derivative at time $t + \alpha h$) by applying the differential equation being solved, to give $\dot{x} = f(t + \alpha h, x_n + \alpha k_1)$.
- This new slope is applied over the whole step h to give an estimate for k_2.

A weighted average of the two k-values is obtained to give x_{n+1}.

The second-order formula represents the Taylor series with all terms removed after the third (terms in h^n where $n \geqslant 3$). It is said therefore to have a truncation error $O(h^3)$ (of the order of h^3). In stating the truncation error in these terms, the assumption is being made that each succeeding term in the Taylor series is smaller than the one before. On the basis of the factorials, this would be true; on the basis of the magnitude of each succeeding derivative, this may or may not be true; and on the basis of the step size h, it is true only if $h < 1$. Generally speaking, however, it is safe to accept that the assumption does hold. The conclusion that arises, therefore, is that the higher the order of a method, the more terms of the Taylor series are represented in the formula, and the smaller the truncation error will be for a given step size.

It is interesting to look qualitatively at the second-order Runge–Kutta method with respect to Figure 6.1. The two k-value estimates represent a determination of two different derivative values, one at the beginning of the step, the other at a point somewhere within the step. These two derivative evaluations represent a rate of change of the first derivative, that is an estimate of the *second-derivative* term in the Taylor series. For a given choice of one parameter (often α), the other parameters may be fixed to give an exact match with the first three terms of the Taylor series.

Since the truncation error is dependent upon the value of h, the step size, this error may be constrained by controlling the particular value of h used during integration. The problem with Runge–Kutta methods is that truncation error is not easy to quantify properly when it is defined as $O(h^2)$, $O(h^3)$, etc. A better estimate of error is obtainable from the next term of the Taylor series, by itself. This estimate is given as the difference between a solution generated by a method of order $m - 1$ and one of order m. Early simulation packages used to have algorithms that generated two parallel solutions just for this reason, but it is more efficient to use the so-called *embedded solution* whereby the $(m - 1)^{\text{th}}$ algorithm is embedded within the mth order one. We can look a little more in depth at this embedded solution idea. What we are looking for is a set of formulae for k-values which are the same for both an $(m - 1)^{\text{th}}$ order method and for an mth order algorithm. After computing the k-values, we have the choice of evaluating x_{n+1} (the value of x at the end of the step) from either a second-order or a third-order formula, as a weighted average of these k-values. Consider the following example for a third-order Runge–Kutta method with a second-order method embedded (Niesse, 1970). The k-values are obtained first of all as follows:

$$\left. \begin{array}{l} k_1 = hf(t, x_n) \\ k_2 = hf(t + 0.5h, x_n + 0.5k_1) \\ k_3 = hf(t + h, x_n - k_1 + 2k_2) \end{array} \right\} \qquad (6.12)$$

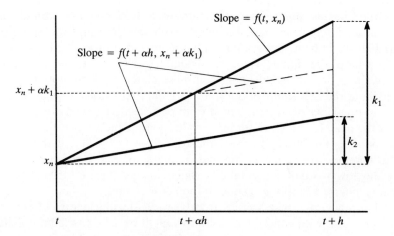

Figure 6.1 Geometric representation of first- and second-order Runge–Kutta integration algorithms.

which gives rise to a well-known third-order Runge–Kutta formula (Ralston, 1956)

$$x_{n+1} = x_n + \frac{1}{6}(k_1 + 4k_2 + k_3) \tag{6.13}$$

Niesse (1970) showed that the same k-values may be used to generate a second-order formula

$$x_{n+1} = x_n + \frac{1}{2}(k_1 + k_3) \tag{6.14}$$

which matched the appropriate terms of the Taylor series. The difference between equations (6.13) and (6.14) constitutes an error estimate for the method. Since the error estimate may be written in absolute terms, it may be written as

$$e_{n+1} = \left| \frac{1}{6}(k_1 + 4k_2 + k_3) - \frac{1}{2}(k_1 + k_3) \right|$$

$$= \left| \frac{1}{3}(k_1 - 2k_2 + k_3) \right| \tag{6.15}$$

Although this method is of low order compared with the fourth-, fifth- and sixth-order Runge–Kutta methods that are commonly in use, it will give quite adequate results if the step size is deliberately kept small for reasons other than error control. This method also has the advantage of calculating k-values at the beginning, midpoint and end of the step, which has implications for its behaviour in the presence of discontinuities, as we shall see later.

Since the starting point for application of a Runge–Kutta formula is the beginning of an integration step, for which both the time and the value of the state variable are known, integration with a Runge–Kutta method can begin at time zero where the initial value is available. This property is known as *self-starting* and is one of the advantages of these methods. Also, nothing needs to be known about the behaviour of the state variable in previous steps at time other than zero, so no values need be retained from past steps. The penalty paid for this freedom is the need for an extra formula evaluation for the error estimate. Strictly speaking, one has the choice to use either the $(m-1)^{\text{th}}$ order or the mth order method to compute the value at the end of the step. Many routines, however, advocate the lower-order computation, which means that a further derivative evaluation is incurred simply to generate the error estimate.

A further advantage that accrues from the self-starting nature of single-step methods is the ease with which step size adjustment may be carried out. The size of an integration step may be changed at will, without any thought about what size prevailed in previous steps. The lack of any past history within an integration step makes every step independent from all others with regard to size.

Finally, in our examination of single-step methods, we should consider briefly the question of *numerical stability*. At this point, it will be sufficient to be aware of the manifestation of numerical instability as it may occur when single-step methods are used. Let us consider the solution of the simple differential equation

$$\dot{x} = \frac{\mathrm{d}x}{\mathrm{d}t} = -10x \quad (x_0 = 1.0) \tag{6.16}$$

by means of the forward Euler formula

$$x_{n+1} = x_n + h\dot{x}_n \tag{6.17}$$

If you look at Figure 6.2, you will see a number of plots superimposed on one set of axes. Apart from the analytical solution shown as a solid line, these are all numerical solutions for the above equation, computed from equation (6.17) with different values of h. The discrete nature of the solutions is emphasized by showing only the solution points. The small step $h = 0.05$ results in solution points quite close to the analytical solution. For $h = 0.1$, all values apart from the initial value are zero, which is a poor (but stable) approximation to exponential decay. The solution for $h = 0.15$ exhibits decaying oscillatory behaviour (therefore still stable), while at $h = 0.2$, the threshold of instability is reached. Any value $h > 0.2$ will result in divergent oscillatory behaviour of the solution transient.

In the normal course of events, a well-behaved single-step method incorporating step length control through error estimation will not exhibit this kind of instability, since the error bounds will usually be tighter than the stability bounds. Implicit methods are more stable than explicit routines. Qualitatively, this is

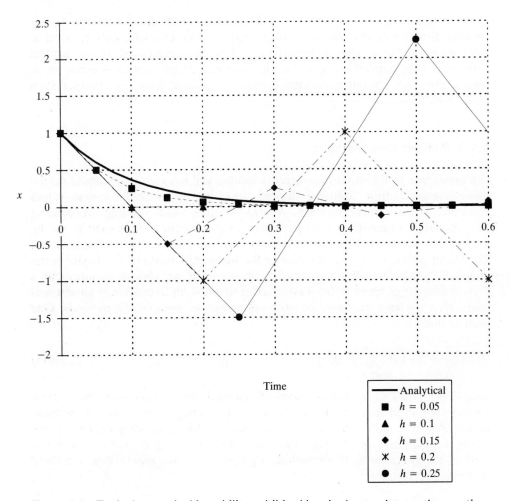

Figure 6.2   Typical numerical instability exhibited by single-step integration routine.

because the very nature of a closed formula (which characterizes the implicit methods) confers a degree of 'smoothing' by virtue of the fact that iteration to convergence ensures that all relationships within the formula are satisfied everywhere at the same time (or some of the relationships in the case of a semi-implicit process). Explicit formulae are extrapolative in nature, and the 'step in the dark' can have dire consequences for stability if the step taken is too large!

Apart from numerical stability, explicit formulae appear to be computationally more efficient, so what other advantages are to be gained by using implicit formulations? One answer lies in the fact that more coefficients can be used in the evaluation for an implicit formula compared with the corresponding explicit form.

If we examine equations (6.8) and (6.9) in the case of the general single-step (Runge–Kutta) formula, we can see that the number of coefficients b_{rs} used at each stage for the k-value is limited to $r-1$ in the summation, for the explicit formula. In the implicit formula, the summation is carried out to m terms, at all stages. We might therefore reasonably expect a greater degree of accuracy from implicit methods.

6.1.3 Multiple Step Methods

We now come to the other main group of numerical integration methods. As the name suggests, these formulae depend upon having available several values calculated at past integration steps. We can therefore adopt a strategy of retaining first-derivative evaluations from past steps and using this information to set up finite-difference approximations for the second- and subsequent-order terms of the Taylor series. In theory, the greater the number of terms of the Taylor series included, the greater the accuracy of the result. If we choose to represent a certain number of terms in the Taylor series by these finite-difference approximations, then we have the basis for what are known as *multiple step methods*. One such formula is

$$x_{n+1} = x_n + h \sum_{j=0}^{k} \beta_j . x'_{n-j} \tag{6.18}$$

This is the well-known *Adams–Bashforth* method, and the terms in the summation are first-derivative evaluations from previous integration steps. The summation represents a finite-difference approximation to higher-order derivatives in the Taylor series. This type of formula may be extended to include past values of x as well as the derivatives. If we include $x_{n-1}, x_{n-2}, \ldots$ we may obtain a general multiple step formula

$$\sum_{j=0}^{k} x_{n-j+1} . \alpha_j - h . \sum_{j=0}^{k} x'_{n-j+1} . \beta_j = 0 \tag{6.19}$$

If $b_0 = 0$ then equation (6.19) is an *explicit* formula, but if $b_0 \neq 0$ then the formula becomes *implicit*.

Because multiple step methods use information from evaluations at past integration steps, they are not self-starting in general. In order to use a multiple step method, you can use a single-step integration scheme for the first few steps in an integration process, before switching to the multiple step method. This applies if you are using a multiple step method of fixed (high) order. There is an alternative to this strategy, and that is to use a *variable-order* multiple step method (such as Gear's algorithm) which starts off as a first-order (Euler) formula, and increases order as more past points become available. This saves employing more than one type of routine in the integration, but the penalty for so doing is that use of a method at low order will require small time steps in order to preserve solution accuracy.

Error estimation is generally good for multiple step methods, for much information is available from the stored values for past integration steps. Without going into detail, we may generally assume that an error estimate is relatively easily obtained from a multiple step integration process, compared with Runge–Kutta methods.

Variation of step size is not a trivial matter when multiple step methods are used. The use of values at past solution points is advantageous when the step size remains constant throughout. However, if the integration step length is to be altered, the grid pattern of past solution values becomes invalid, and interpolation techniques are required to construct a new pattern appropriate to the new step size, before the calculation can proceed further.

Although multiple step methods are capable of giving good error estimation, a particular type of instability may arise in their use. The general multiple step formula, equation (6.19), is a finite-difference equation, and for $k > 1$ will have more than one solution, only one of which approximates the true solution of the original differential equation. A multiple step method is stable if the spurious solutions do not swamp the desired true solution. Truncation error control through step size adjustment should automatically ensure stability in any reasonable multiple step method.

We have seen that both single- and multiple step integration algorithms may take on both explicit and implicit forms. The general multiple step formula, equation (6.19), has at least one coefficient more in the implicit formula than in the explicit one. By the same token as with single-step methods, the extra information contained within an implicit formula should mean that greater solution accuracy is potentially available than is the case with the corresponding explicit algorithm.

Following the analogy of semi-implicit single-step methods, we may extend similar 'cost-saving' measures to working with implicit multiple step formulae to accelerate convergence to a solution. This is accomplished by using an explicit formula as a *predictor* to obtain a fairly close approximation to the eventual solution. This is used as the input to a closed formula *corrector*, an implicit algorithm that converges to a better solution in usually two or three iterations. Let us consider a simple *predictor–corrector* pair.

Predictor: $\bar{x}_{n+1} = x_n + hx'_n$ (6.20)

This is the *forward Euler formula* with truncation error $O(h_2)$.

Corrector: $x_{n+1} = x_n + \dfrac{h}{2}(\bar{x}'_{n+1} + x'_n)$ (6.21)

This is the *trapezoidal rule* with truncation error $O(h^3)$. In this algorithm, $\bar{x}_{n+1}$ is the 'approximate' solution determined by the predictor formula. The corresponding derivative $\bar{x}'_{n+1}$ is computed from the differential equation, and used in the corrector formula to obtain an improved solution value. This is then applied again within the corrector formula and the process repeated until two successive iterations produce the same solution to within a degree of tolerance. One application of the predictor formula is followed by as many applications of the corrector as are necessary to achieve convergence. Even for high-order methods, as few as three derivative evaluations may be needed (one pass through the predictor and two passes through the corrector). Compare this with commonly used fourth-order Runge–Kutta methods with error control, which need five derivative evaluations.

6.2 STEP SIZE CONTROL FOR NUMERICAL INTEGRATION

In numerical integration, two options may be exercised with regard to integration step size. The integration step length, once determined, may be kept constant throughout the whole of a simulation run, or the step size may be allowed to vary in line with the behaviour of the solution transients. Let us consider both of these options.

6.2.1 Fixed Step Integration – useful for some applications, but beware!

Many simulationists are of the opinion that fixed step integration is quite adequate for most simulation studies. This may be true provided that the step length to be used is properly determined at the outset, in terms of the maximum value it can have for the preservation of accuracy and numerical stability throughout a simulation run. While this may be perfectly satisfactory if the maximum value can indeed be properly assessed, it may be inefficient if the step size could be larger over a significant portion of the simulation time span. The greatest difficulty in using fixed step length integration is being able to determine what step size will be suitable at all points during the solution, to fulfil the accuracy criteria that have been set, and without incurring numerical instability.

Fixed step integration does have advantages for one application, namely for real-time simulation. If a simulation has to run in real time, because it is interfaced with real control hardware, or for some other reason, then fixed step integration can facilitate the timing operations involved. It is much easier to obtain real-time operation if the execution time for the integration algorithm is constant by virtue of a fixed step procedure. Variable step integration is not nearly so predictable, as we shall see, and therefore its application to real-time operation imposes greater difficulties in implementation.

6.2.2 Why Vary the Step Size?

Under rapidly changing transient conditions, derivatives (and hence truncation errors) will be large, so the step size needs to be correspondingly small. Under quiescent conditions of near steady state, truncation errors may be small enough to allow a considerable increase in step size without jeopardizing solution accuracy. Although a computation overhead will be incurred for error estimation, step size variation may well be more efficient in the long run, particularly if step length is always maximized. At the very least, it should mean that solution errors are kept within bounds, whatever the behaviour of solution transients may be. A further spin-off will arise from the use of variable step integration, if unpredictable discontinuities appear within the solution.

6.2.3 Error Control in Numerical Integration

We now come to the practical implications of error control during numerical integration. In our simplified discussion of the numerical analysis of integration methods, we have come to terms with the fact that numerical methods for integration do not produce a perfect error-free solution. The means whereby this might have been achieved – the Taylor series – has been approximated by routines that attempt to match only a finite number of terms of the series, hence the truncation error that arises during integration.

In putting error control into practice, we need to specify the error criteria that we wish to have enforced during a simulation run. Then we have to devise some means of implementing these criteria and translating the specifications into control of the error estimates produced by the integration algorithm. Before we decide what form our practical error criteria should take, we may consider, first of all, the means whereby step size is adjusted to keep errors within bounds. Two methods of adjustment are in common use.

6.2.3.1 Binary adjustment of step size

The first method of step size adjustment is the so-called *binary* method, whereby the following procedure is carried out.

- Apply the integration routine to the differential equation over the chosen step. Compute the error estimate for that step.

- If the error estimate is larger than the error criterion will allow (we may call this the *tolerance*), reset all parameters to the beginning of that same step, and try again with *half the original step size*.
- If the error estimate is within bounds, test to see if it is within a small fraction (often 1/32) of the tolerance.
- If the error estimate is small enough, reset all parameters to the beginning of that same step and try again with *double the original step size*.
- Otherwise, update the solution for that step and proceed with the next step.
- If the process is being applied to more than one differential equation, the maximum error estimate found, or a 'norm' of the vector of error estimates, is used to make the decisions in both cases.

Certain difficulties may arise out of the application of this procedure. Unless precautions are taken, it is possible to get into a situation where one step doubling is followed immediately by a step halving, and as a consequence the integration process makes no headway. Even if this extreme situation is not encountered, it is a wise precaution after a step halving to force the completion of several steps at the new length, before once again permitting doubling to take place. Because of the step rejection, the binary step adjustment process is rather inefficient. It is, however, a secure method of working.

6.2.3.2 Continuous adjustment

The other method in common use is the method of continuous adjustment. This is a slight misnomer, for it is usual to reject a step that fails the error test, but the step size adjustment is continuous in that it is adjusted at each step according to the error estimate found at the last step, whether or not the tolerance was exceeded. The formula for this adjustment (Hull *et al.*, 1972) is given as

$$h_{\text{new}} = \min\left[h_{\text{max}}, 0.9 h_{\text{old}} \left(\frac{\text{tolerance}}{\text{error estimate}} \right)^{1/m} \right] \tag{6.22}$$

where m is the order of the method, and h_{max} is a maximum step size defined by other constraints. This formula is applied in the following way.

- Apply the integration routine to the differential equations being solved. Compute the error estimates for all variables, and use either the maximum or a 'norm' to test against the tolerance.
- If the error test fails, reset the parameters to retry the same step.
- If the error test succeeds, update the solution.
- Compute a new step size from equation (6.22) using the error estimate found from the previous integration, whether successful or not. Proceed with the new step.

This method has the advantage of being able to advance the simulation time at each step (apart from cases of step rejection), so it is more efficient than the binary process.

6.2.3.3 Error criteria as set by the user

In continuous-variable simulation using digital solution methods, variable step numerical integration requires the specification of one or more error criteria. These are combined in various ways to give a tolerance against which the error estimates of the algorithm are judged. In the absence of any action by the user, default values are set within simulation packages.

Two different accuracy criteria are usually necessary for integration. Both are set as decimal fractions to express the permitted size of an error estimate indicator itself as a fraction:

- *Absolute accuracy* is a fixed figure of merit. When a differential equation solution variable (state variable) has magnitude less than one, the error estimate is set against this fixed fractional value.
- *Relative accuracy* expressed as a fraction of the solution magnitude is used when the state variable assumes magnitudes greater than one. It would be unreasonable under these circumstances to expect proper comparison of an error estimate with an absolute value.

If we have a state variable x, an absolute accuracy A and a relative accuracy R, then there are different ways in which we may implement an error test for the numerical integration of x with an error estimate E:

- Separate implementation of accuracy criteria:

  If $|x| < 1.0$, then compare E with A (tolerance $= A$)
  If $|x| > 1.0$, then compare E with $R.x$ (tolerance $= R.x$)

- Sum of the two accuracy criteria:

  Compare E with $A + R.x$ (tolerance $= A + R.x$).

These allow separate specification of A and R, which may be useful in certain circumstances. If, however, there is no objection to having the same value set for both A and R (which we shall call *accuracy*), then the following formula can be quite efficient in its application:

- Tolerance $=$ accuracy $\times \max(|x|, 1.0)$.

  Compare E with tolerance.

In the above applications, if we are solving several equations instead of just one, then we may replace x and E with 'norms' of the respective vectors, or we may evaluate the relationships for each equation separately.

6.3 THE NUMERICAL ANALYSIS OF 'STIFFNESS'

You will remember from Section 5.6.4 that the concept of 'stiffness' was introduced with regard to the digital solution of differential equations. We are now in a position to examine the numerical analysis of 'stiffness' and to see how the problem may be alleviated.

The basis of all methods for the solution of sets of differential equations with a wide range of time constants (or eigenvalues, if you prefer it that way) is the use of implicit integration formulae.

Let us consider why this should be so, with reference (once again) to the solution of a simple first-order differential equation

$$\dot{x} = f(x) = -\frac{1}{\tau}x \tag{6.23}$$

Equation (6.16) was a particular example of this with $\tau = 0.1$. The solution may be expressed analytically as

$$x = x_0 . e^{-t/\tau} \tag{6.24}$$

and is stable because $x \to 0$ as $t \to \infty$.

If a simple explicit numerical integration method (e.g. the forward Euler formula) is applied to the solution of this equation, then

$$x_{n+1} = x_n + hf(x_n)$$

$$= x_n - \frac{h}{\tau}x_n$$

$$= \left(1 - \frac{h}{\tau}\right)x_n \tag{6.25}$$

If the step length h is held constant throughout, then we may write

$$x_{n+1} = \left(1 - \frac{h}{\tau}\right)^{n+1} . x_0 \tag{6.26}$$

The minimum requirement for the numerical solution to follow the analytical solution is that $x_{n+1} \to 0$ as $n \to \infty$. The condition required for this to happen is

$$\left|1 - \frac{h}{\tau}\right| < 1 \tag{6.27}$$

from which

$$h/\tau < 2 \tag{6.28}$$

and

$$h < 2 . \tau \tag{6.29}$$

If this condition is not met, then the classical single-step instability (non-decaying and oscillatory transients) is the result. Figure 6.2 gave us an illustration of this phenomenon in the case where the time constant for the equation was given as $\tau = 0.1$. A step size, $h = 0.2$, defined the limiting stability situation, for which the solution consisted of undamped oscillations which neither decayed nor enlarged. Equations with very short time constants (large eigenvalues) will limit the maximum step length that may be used, while those with long time constants may require long solution times for the full transient responses to be seen.

Let us now consider the use of a simple implicit integration scheme (e.g. the backward Euler formula). We may express this as

$$x_{n+1} = x_n + hf(x_{n+1}) \tag{6.30}$$

where the derivative is a function of the value of x at the *end* of the integration step. If we apply this to equation (6.23), we may write

$$x_{n+1} = x_n - \frac{h}{\tau} x_{n+1} \tag{6.31}$$

Solving for x_{n+1}, we obtain the equation

$$x_{n+1} = \frac{1}{(1 + h/\tau)} x_n \tag{6.32}$$

and if we assume constant step length throughout, we may write

$$x_{n+1} = \frac{1}{(1 + h/\tau)^{n+1}} x_0 \tag{6.33}$$

For all values of h, $x_{n+1} \rightarrow 0$ as $n \rightarrow \infty$, so the integration scheme is stable and suitable for differential equations with short time constants.

Equation (6.30) is a closed formula, that is the solution variable x_{n+1} appears on both sides of the equation. On the assumption that the derivative value $f(x_{n+1})$ is non-trivial, we cannot solve by direct substitution, and some iterative method of solution becomes necessary. Let us write equation (6.30) in a different way:

$$x_{n+1} = x_n + hf(x_{n+1})$$

$$= g(x_{n+1}) \tag{6.34}$$

We may solve for x_{n+1} iteratively using, for example, the Newton–Raphson method

$$x_{n+1}^{s+1} = x_{n+1}^s - \frac{g(x_{n+1})}{g'(x_{n+1})} \tag{6.35}$$

where g' denotes derivative with respect to x (d/dx) and the superscript s refers to

the number of iterations of the Newton formula. We may therefore write
equation (6.35) in a different way:

$$x_{n+1}^{s+1} = x_{n+1}^s - g(x_{n+1}) \left(\frac{\mathrm{d}}{\mathrm{d}x} g(x_{n+1}) \right)^{-1} \tag{6.36}$$

When an implicit method like this one is applied to the solution of sets of
differential equations, the single-derivative evaluation of the Newton–Raphson
method is replaced by the evaluation of the inverse of a Jacobian matrix of partial
derivatives of the form $\partial g(x_i)/\partial x_j$. If you use one of the well-known implicit
solution schemes for 'stiff' sets of differential equations, such as the method of
Gear (1971), you will come across many references to the evaluation of the
Jacobian matrix and its inverse. Computationally, this evaluation involves a
considerable overhead, but in the nature of the solution method, it is not usually
necessary to calculate it every integration step. Since Gear's method is a
predictor–corrector, non-convergence of the corrector formula is a sign that the
Jacobian matrix needs to be updated.

6.4 WHICH DO WE SOLVE, ODEs OR DAEs?

Before any answer can be given to this question, it is necessary for us to be clear
what is meant by the abbreviations ODEs and DAEs in the context of simulation.
The term *ODE* as we have already seen, means *ordinary differential equation* and
in the context of simulation, solving ODEs means computing the derivative
expressions of a set of first-order differential equations, and using this information
to update the values of the corresponding state variables. To recapitulate what we
covered at length in Section 5.6.1, a circular sequence of computations takes
place, comparable with the circuitry of an analog computer model. We start with
the current values of the state variables, work through the solution of all the
algebraic equations in strict computational order, and end up with the model
derivatives. These are passed to the integration algorithm and the state variables
are updated. In this way we achieve a quasi-parallel solution to the total set of
equations.

The term *DAEs* is a shorthand way of saying *differential and algebraic
equations*. In the context of simulation, solving DAEs means treating the
complete model equation set as a single entity as far as the integration algorithm
is concerned. The rationale for doing this is the use of implicit methods of
solution. In the solution of ODEs, the use of an implicit method requires the
iteration of a closed formula containing only the state variables and their various
derivative manifestations. There seems therefore to be no reason at all why the
iteration scheme should not be expanded to take in all the algebraic equations as
well. In this way, a more truly parallel solution would be obtained to the set of
differential and algebraic equations (DAEs).

Let us consider equation (6.2), this time written in vector form to represent a *set* of differential equations

$$\mathbf{x}'(t) = \mathbf{f}(t, \mathbf{x}(t)) \tag{6.37}$$

with initial values

$$\mathbf{x}(0) = \mathbf{x}_0$$

where the boldface type indicates vector notation. We may express this in our step-wise notation as

$$\mathbf{x}'_n = f(t_n, \mathbf{x}_n) \tag{6.38}$$

If we now apply the backward Euler formula to the solution of this set of equations, and substitute equation (6.38), we have

$$\mathbf{x}_{n+1} = \mathbf{x}_n + h\mathbf{x}'_{n+1}$$

$$= \mathbf{x}_n + h\mathbf{f}(t_{n+1}, \mathbf{x}_{n+1}) \tag{6.39}$$

Also, from equation (6.39), we may write

$$\mathbf{x}_{n+1} - \mathbf{x}_n = h\mathbf{x}'_{n+1} \tag{6.40}$$

$$\therefore (\mathbf{x}_{n+1} - \mathbf{x}_n)/h = \mathbf{x}'_{n+1} \tag{6.41}$$

Since we may define the step size as

$$h = t_{n+1} - t_n \tag{6.42}$$

we may write equation (6.41) as

$$\mathbf{x}'_{n+1} = \frac{\mathbf{x}_{n+1} - \mathbf{x}_n}{t_{n+1} - t_n} \tag{6.43}$$

If we solve the set of equations represented by equation (6.37) implicitly, then we may represent the equation set in implicit form as

$$\mathbf{F}[t, \mathbf{x}(t), \mathbf{x}'(t)] = 0 \tag{6.44}$$

Again in step-wise notation, we may write this as

$$\mathbf{F}[t_{n+1}, \mathbf{x}_{n+1}, \mathbf{x}'_{n+1}] = 0 \tag{6.45}$$

If we substitute equation (6.43), we have

$$\mathbf{F}\left[t_{n+1}, \mathbf{x}_{n+1}, \frac{\mathbf{x}_{n+1} - \mathbf{x}_n}{t_{n+1} - t_n}\right] = 0 \tag{6.46}$$

This would constitute a set of finite-difference equations that could be iterated to a solution. If we now go back to equation (6.37), but instead of a set of differential equations only, we have a set of differential and algebraic equations of the form

$$\mathbf{x}'(t) = \mathbf{f}(t, \mathbf{x}(t), \mathbf{y}(t)) \tag{6.47}$$

with initial values

$$\mathbf{x}(0) = \mathbf{x}_0$$

and

$$\mathbf{y}(t) = \mathbf{g}(t, \mathbf{x}(t), \mathbf{y}(t)) \tag{6.48}$$

In the same way as for equation (6.44), we may express this pair of equations in implicit form as

$$\mathscr{F}[t, \mathbf{x}(t), \mathbf{x}'(t), \mathbf{y}(t)] = 0 \tag{6.49}$$

At the $(n+1)^{\text{th}}$ step, this will take the form

$$\mathscr{F}[t_{n+1}, \mathbf{x}_{n+1}, \mathbf{x}'_{n+1}, \mathbf{y}_{n+1}] = 0 \tag{6.50}$$

We may substitute equation (6.43) to give us the finite-difference formula

$$\mathscr{F}\left[t_{n+1}, \mathbf{x}_{n+1}, \frac{\mathbf{x}_{n+1} - \mathbf{x}_n}{t_{n+1} - t_n}, \mathbf{y}_{n+1}\right] = 0 \tag{6.51}$$

which may be iterated to a solution in $\mathbf{x}$ in the same way as equation (6.46). This represents the application of the backward Euler method to the implicit solution of a set of differential and algebraic equations (DAEs).

If all the model equations were to be written in the form of equation (6.51), the right-hand sides of the expressions would be *residuals* (rather than zero in each case). Iteration would proceed until the residuals had all reached values close to zero to within a tolerance. The 'norm' of the residuals is a useful figure of merit; when that is close to zero, all the residuals must be very small in magnitude as well.

One difficulty that arises with the application of the implicit solution to DAEs rather than ODEs is initialization of the system of equations at time zero (or some other starting time value). At time $t = 0$ (or $t = t_0$, if preferred), equation (6.50) becomes

$$\mathscr{F}[t_0, \mathbf{x}_0, \mathbf{x}'_0, \mathbf{y}_0] = 0 \tag{6.52}$$

This means that, if we have included the algebraic equations of the system in the implicit formula equation (6.44) to give equation (6.49), we then have to assume that equation (6.49) is as true at time $t = 0$ (t_0) as it is at any other time. For initialization of the complete set of DAEs, therefore, iteration is required to solve equation (6.52) for *all* initial values, whether state or algebraic variables. This also includes initial values for the derivatives, which, as we have already seen in Section 5.3.1, are really algebraic variables. In many instances this initialization process can be difficult to achieve. Unlike iteration within the integration algorithm, which starts from an explicit solution (a good approximation), initialization for a DAE solution may need much help from the user in specifying a suitable starting point.

A considerable advantage in the implicit solution of DAEs is the fact that it is not necessary to re-arrange equations to extract the solution variable onto the left-hand side. In a simulation package implementing this technique, the facility may exist whereby model equations may be coded in the form in which they were specified for the model. This is particularly advantageous if the model equations are difficult to manipulate into an analog computer type of solution format, each with a single solution variable on the left-hand side of the equation.

By the same token, there are dangers in using this implicit DAE solution technique. The discipline of having to consider carefully the equations specified for a simulation model, particularly in regard to the solution variable for each, can help one to diagnose and avoid incorrect model specifications. DAE solution will probably diagnose the presence of poor model structure (through the properties of the Jacobian matrix), but it will not necessarily be readily apparent where the fault lies.

A further advantage of the DAE implicit solution approach is obviation of the need either to have all model equation coding statements in the correct computational sequence, or to have to employ a sorting algorithm to order the statements properly. Also, the phenomenon of algebraic loop occurrence is of no consequence in DAE solution.

Once again, it is questionable whether this relaxation of constraints is altogether beneficial in the development of a good simulation model. The discipline required to order the coding statements, or at least the error messages output by a sorting algorithm, can concentrate the mind very well on giving proper thought to the equations being used in a model description.

One problem that comes to light in DAE solution is the question of *solution index*. By virtue of the iterative nature of DAE solution, there arises the temptation to include extra equations in a model definition, which ODE solution methods cannot handle directly. For example, a constraint may be put upon the behaviour of one of the solution variables in an equation set, such that it is expected to behave in a fixed way determined by an additional equation. An ODE solution would require repeated solution to satisfy such a constraint, and it is natural to expect DAE solution to be able to include extra equations of this nature, without any problem.

Let us consider a simple example of a system that is not an index problem, and see how we might turn it into one. Figure 6.3 shows a circuit diagram for a simple *LRC* circuit. Application of an input voltage e_1 results in an output voltage e_2 according to the relationship

$$e_1 = L\frac{di}{dt} + R_1 i + e_2 \qquad (6.53)$$

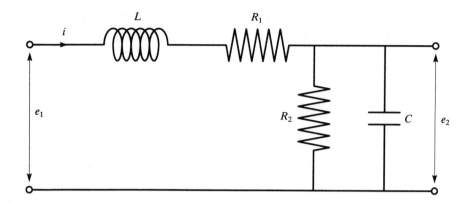

Figure 6.3   Diagram of a simple *LRC* circuit.

where

    i is current
    L is inductance
    R is electrical resistance

If we re-arrange equation (6.53), we obtain the ODE

$$L\frac{di}{dt} = e_1 - R_1 i - e_2 \tag{6.54}$$

Also by application of Kirchhoff's Law

$$i = C\frac{de_2}{dt} + \frac{e_2}{R_2} \tag{6.55}$$

where

    C is capacitance

We may re-arrange equation (6.55) to obtain

$$C\frac{de_2}{dt} = i - \frac{e_2}{R_2} \tag{6.56}$$

Equations (6.54) and (6.56) are sufficient to describe completely the behaviour of the circuit in terms of its response to changes in the input voltage e_1. The solution index for this problem is therefore zero, since the system model is minimally realized in terms of ODEs.

    Loosely speaking, the solution index is the number of differentiations that have to be applied to a set of model equations in order to obtain a model description in terms of ODEs only, with no algebraic equations present. Let us therefore now consider an alteration to our circuit model that increases the solution index. As so often happens, equations (6.54) and (6.56) constitute a second-order system which is capable of oscillatory behaviour if the circuit components are chosen appropriately for this to be the case. If we differentiate equation (6.55) to obtain

$$\frac{di}{dt} = C\frac{d^2e_2}{dt^2} + \frac{1}{R_2}\frac{de_2}{dt} \tag{6.57}$$

we may then substitute equations (6.55) and (6.57) into equation (6.53) to give

$$e_1 = LC\frac{d^2e_2}{dt^2} + \left[\frac{L}{R_2} + R_1C\right]\frac{de_2}{dt} + \left[\frac{R_1 + R_2}{R_2}\right]e_2 \tag{6.58}$$

which is a second-order differential equation. *LRC* circuits may be designed to 'ring', that is give rise to a highly oscillatory output in response to input voltage changes. On the other hand, it may be considered desirable for the output voltage e_2 to exhibit more orderly behaviour. The problem then arises as to what form the input voltage e_1 should have to bring this about. Let us introduce an algebraic equation to describe the required behaviour for e_2. If we constrain e_2 to be a ramp function with respect to time, starting from a value of zero, then we may write the equation

$$e_2 = k_1t \tag{6.59}$$

Equations (6.54), (6.56) and (6.59) constitute the equations for what is known as an *implied control model*. An equation describes the required behaviour of a controlled variable, the assumption being implicit that the rest of the model equations will fall into line to produce the desired result. If such a set of equations can be solved, then this is precisely what does happen.

Having forced the constraint of equation (6.59) upon the system, we may no longer assume that arbitrary values may be given to the input voltage e_1, or indeed to any of the initial values for the variables and their derivatives. The problem of determining a consistent set of initial values that will yield a solution to all three equations (6.54), (6.56) and (6.59) is not particularly difficult to solve for this simple system, but for large and complex sets of equations, determination of a correct and consistent set of initial values can be very troublesome. Let us consider this example to see how we may go about this task.

If we differentiate equation (6.59) to give

$$\frac{de_2}{dt} = k_1 \tag{6.60}$$

and substitute equation (6.60) into equation (6.55), we then obtain

$$i = k_1C + \frac{e_2}{R_2} \tag{6.61}$$

We may then differentiate equation (6.61) to give

$$\frac{di}{dt} = \frac{1}{R_2}\frac{de_2}{dt} \tag{6.62}$$

By substituting equation (6.60), we eliminate the derivative on the right-hand side of the equation to obtain

$$\frac{di}{dt} = \frac{k_1}{R_2} \tag{6.63}$$

Using equation (6.53), we may then obtain an expression for e_1 as

$$e_1 = \frac{Lk_1}{R_2} + R_1 i + e_2 \tag{6.64}$$

which describes the behaviour of input voltage e_1 that is required if the output voltage e_2 is to be generated in accordance with equation (6.59). Equations (6.60), (6.63) and (6.64) constitute a new version of the original implied control model, which is now of index one (one differentiation of equation (6.64) reduces the model to a set of ODEs). The new model definition is easily soluble by ODE methods in the normal way. Note that the initial conditions are now constrained as follows:

$$e_2(0) = 0 \qquad \text{(from equation (6.59))} \tag{6.65}$$

$$i(0) = k_1 C \qquad \text{(from equation (6.61))} \tag{6.66}$$

$$e_1(0) = \frac{Lk_1}{R_2} + R_1 k_1 C \quad \text{(from equation (6.64))} \tag{6.67}$$

$$\left.\frac{de_2}{dt}\right|_0 = k_1 \qquad \text{(from equation (6.60))} \tag{6.68}$$

$$\left.\frac{di}{dt}\right|_0 = \frac{k_1}{R_2} \qquad \text{(from equation (6.63))} \tag{6.69}$$

This causes no problems for an ODE type of solution, for the state variable initial values are specified in the normal way. The input voltage e_1 and the derivatives are defined by the model equations in the usual manner. It is only when implicit DAE solution is carried out that the whole set of equations must be properly initialized in terms of all initial values together as specified above.

6.5 REFERENCES

CAILLAUD, J B and PADMANABHAN, L (1971) 'An improved semi-implicit Runge–Kutta method for stiff systems', *Chem. Eng. J.*, vol. 2, pp. 227–32.

GEAR, C W (1971) 'The automatic integration of ordinary differential equations', *Commun. ACM*, vol. 14, pp. 176–9.

HULL, T E, ENRIGHT, W H, FELLEN, B M and SEDGWICK, A E (1972) 'Comparing numerical methods for ordinary differential equations', *SIAM J. Numer. Anal.*, vol. 9, pp. 603–37.

NIESSE, D H (1970) 'Technical comment on low-order variable step Runge–Kutta methods', *Simulation*, vol. 14, pp. 93–4.

RALSTON, A (1956) *A first course in numerical analysis*, International Student Edition, Tokyo, McGraw-Hill Kogakusha, p. 199.

ROSENBROCK, H H (1963) 'Some general implicit processes for the numerical solution of differential equations', *Comput. J.*, vol. 5, pp. 329–30.

7

Discontinuities in Digital Simulation

In Section 5.6.3, we touched upon the subject of discontinuities as one of the problem areas of simulation where digital computation encounters considerable difficulty, but to which analog computation was immune. This can cause such severe problems in slowing down numerical integration that it will pay us to delve more deeply into the causes and effects, in order to try to alleviate the problem. Having covered something of the numerical analysis of digital integration, we are now in a position to examine the effects of discontinuities from this point of view.

7.1 THE EFFECTS OF DISCONTINUITIES ON NUMERICAL INTEGRATION

Our look at the numerical analysis of digital integration has shown us that time advance is effected basically through the ability of an integration algorithm to make a prediction concerning the value that a state variable should have at the end of a time step. This prediction is based on an analysis of past results (multiple step methods) or various estimates carried out within the current time step (single-step methods). A discontinuity that affects the direction taken by the state variable derivative is likely to upset the prediction. Let us examine the way in which a discontinuity causes problems by slowing down the speed of execution, for both single- and multiple step integration methods.

7.1.1 Single-step Integration Algorithms

Figure 7.1 shows a simple diagram of potentially the worst kind of discontinuity that may beset an integration algorithm. It occurs in the evaluation of a derivative (Figure 7.1(a)), and results in an abrupt change in the direction being taken by the solution transient (see Figure 7.1(b)).

Let us recall the second-order Runge–Kutta integration method embedded within a third-order routine (equations (6.12) to (6.14)). The second-order

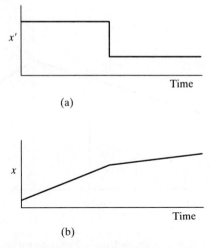

Figure 7.1 (a) Discontinuity in the derivative of a state variable; (b) solution transient for the same state variable.

method computes an estimate which is the average of the first and last of the three k-values computed in equation (6.12). Let us also consider the application of this routine to a problem with a discontinuity in the solution. Figure 7.2 shows part of the transient where one step in the integration process includes the discontinuity at a point just after the half-way mark (although it might just as easily be anywhere else within the step). Calculation of k_1 produces a value for $x_{n+1/2}$ $(x_n + 0.5k_1)$ from which k_2 is then computed. So far, so good, but since k_3 is computed as a function of k_1 and k_2 and as a function of time at the end of the step $(t + h)$ (i.e. after the discontinuity), that is when our troubles begin. k_3 will be seen to be very different from k_1 and k_2, on account of the effects of the discontinuity, and this discrepancy is seen as a large truncation error. The step is rejected, and the step size reduced in accordance with the observed error estimate. If the resulting new step stops short of the discontinuity, that step is completed in the normal way. If the next step after that straddles the discontinuity, the step reduction process is repeated. Eventually, when the step size is small enough for the observed truncation error to fall within tolerance, the discontinuity will be passed, and step lengths can return to normal.

7.1.2 Multiple Step Methods

Let us now consider what happens when we try to integrate over a discontinuity, using a multiple step integration scheme. Figure 7.3 shows part of the transient from such a process, including the discontinuity. In multiple step integration, the results from previous steps will predict a likely direction to be taken in the next

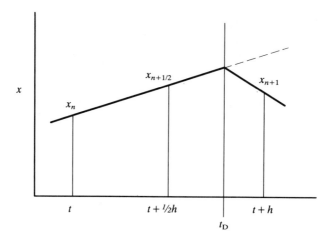

Figure 7.2   Discontinuity occurrence within second-order Runge–Kutta integration step.

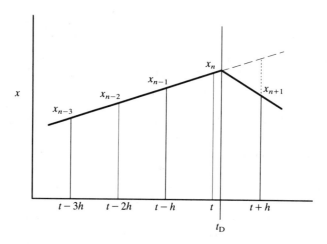

Figure 7.3   Discontinuity occurrence within next step of multiple step integration algorithm.

step. If a derivative discontinuity falls within this next step, the direction taken changes radically from that expected, and the result is seen as an excessively large truncation error by the integration algorithm. The algorithm subdivides the step in an attempt to reduce this error, and will do so successively until the greatly reduced step size ensures close proximity to the discontinuity, and the truncation error has become acceptably small for the discontinuity to be negotiated successfully.

7.1.3 Example: Sawtooth Generation

A standard problem that used to be given to anyone learning the 'art' of parallel logic analog computation was that of generating a sawtooth waveform, using one integrator and a complement of logic and interface devices to change the direction of integration as required.

If x is the solution (state) variable whose value generates the waveform, MAX is the peak value, MIN is the trough value, A is the positive slope, $-B$ is the negative slope and L is a latch that determines (and holds) the direction of integration, then we may define the model for sawtooth generation as follows:

$$\frac{dx}{dt} = A \qquad (MIN \leqslant x \leqslant MAX) \wedge (L = 1) \tag{7.1}$$

$$\frac{dx}{dt} = -B \qquad (MIN \leqslant x \leqslant MAX) \wedge (L = -1) \tag{7.2}$$

$$L = 1 \qquad (x \leqslant MIN) \wedge (L = -1) \tag{7.3}$$

$$L = -1 \qquad (x \geqslant MAX) \wedge (L = 1) \tag{7.4}$$

If we choose $A = 3$, $B = 1.5$, MIN $= 0$ and MAX $= 1$, then the peaks of the sawtooth will occur every $(N + 0.3333)$ time units $(N = 0, 1, 2, \ldots)$, and the troughs will occur every N time units $(N = 0, 1, 2, \ldots)$. In this way the peaks should occur at points that are not coincident with normal output times, although the troughs are likely to occur at output points, since the sawtooth period is exactly 1 time unit.

The first digital implementation of the sawtooth model is written in the ACSL language (SAWTL.CSL and SAWTL.CMD listed in Appendix A.1). Figure 7.4 shows the post-run plot of the sawtooth waveform and slope shown on two separate strip plots. The sawtooth slope should exhibit a square waveform, but this is certainly not the case. The sawtooth waveform itself should peak at a value of 1.0, but falls far short of this for all peaks. However, the troughs do occur at $x = 0$. An explanation for the poor performance is given as follows:

- Latch operation occurs too soon on the upstroke, possibly because one k-evaluation in the Runge–Kutta algorithm detected a value of x greater than 1, even though the eventual result was less.
- The sawtooth waveform peaks do not lie on output points; hence they would be passed over without being properly registered. This could account for the non-square waveform for the slope, since the changeover of slope value would only be registered between two output points (0.1 time unit apart).
- If an integration step is forced to end at an output point that coincides with a trough, the latching operation is more likely to operate accurately and change sign at $x = 0$.

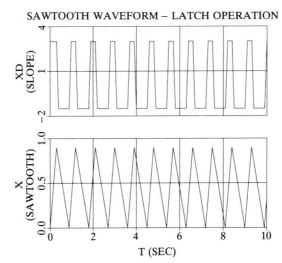

SAWTOOTH WAVEFORM – LATCH OPERATION

Figure 7.4   Post-run ACSL plots of sawtooth waveform and slope, latch operation,
CINT = 0.1.

- The errors in the peak changeover, however, cause severe timing errors to occur such that the period is no longer 1.0.

A second run was carried out with a communication interval of 0.001 time unit, the results of which are shown plotted post-run in Figure 7.5. The performance of the waveform generation is greatly improved, since practically all the 'aliasing' of the up-peaks has been removed, as well as that for the square waveform of the slope. Also, the sawtooth period is very close to 1 time unit as predicted, with little error build-up taking place over the timescale of the simulation. The penalty for this improvement is a much shorter step length brought about by the small communication interval.

A second implementation of the sawtooth model was done using MATSIM. The two model files (SAWTUNIN.M and SAWTUNB.M) are listed in Appendix A.2. Operation of the model is a little different from the ACSL implementation, in that use is made of the extra facility that is included in MATSIM for carrying out processing at the end of a completed integration step (see Appendix A.2 for details). Lines 23–5 of the listing for SAWTUNB.M show that the latch is only operated if the sawtooth amplitude has reached a peak or trough *at the end of a completed step*. Slope changes for the waveform may be carried out on a temporary basis, to register the presence of a discontinuity within the current integration step. This processing is shown in lines 12–22 of the model listing. Latch operation then effects a permanent changeover of slope until the next discontinuity is reached.

Figure 7.6 shows a run-time plot from MATSIM. Since output is at every integration step, there is no aliasing of the waveform, and it appears reasonably

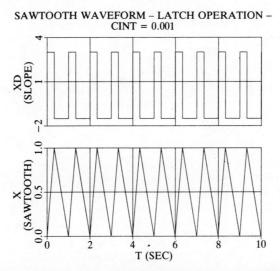

Figure 7.5 Post-run ACSL plots of sawtooth waveform and slope, latch operation, CINT = 0.001.

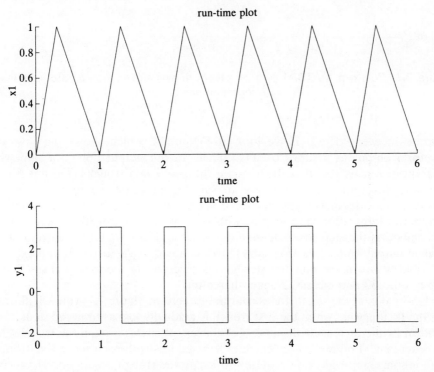

Figure 7.6 Post-run MATSIM plots of sawtooth waveform and slope, latch operation.

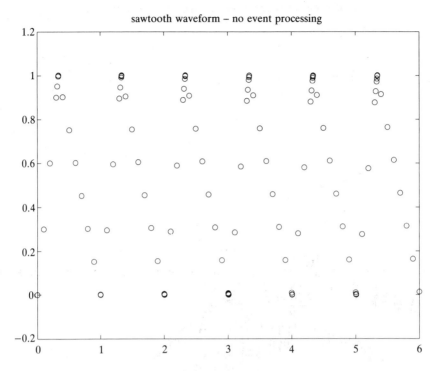

Figure 7.7   Post-run MATSIM plot of sawtooth waveform, point output at every integration step.

accurate in every detail. If we look at Figure 7.7 which shows the sawtooth waveform output as a series of small circles at the output points, a somewhat different picture is revealed. In between the peaks and troughs, the integration step is quite regular and of a size to suit the required integration accuracy. However, as integration approaches a waveform peak, there is a noticeable decrease in integration step length, with steps very close together near a peak. This embodies the step size reduction discussed in Section 7.1.1. A portion of the detailed output obtainable from MATSIM is shown in Table 7.1. If you examine this table, you can see how the step size is successively reduced near a value of time $t = 0.3333$ and increased again thereafter.

If you look at the small circles near the troughs in Figure 7.7, you will also see that the accuracy shown in the first trough is gradually lost as time proceeds. This is shown by the increasing dispersion of clusters of circles around the troughs. As the programmed output point coincides less and less with each successive trough, there is some step reduction required to locate the trough accurately. If you run MATSIM to produce a run-time plot, the delays caused by step reduction will be quite evident.

7.2 ALLEVIATING DISCONTINUITY EFFECTS

Ideally we should like the occurrence of a discontinuity to be totally transparent to a numerical integration algorithm in the same way that we perceive with analog integration. Fortunately, we can come quite close to achieving this ideal with some integration methods and certain types of discontinuity.

7.2.1 Single-step Methods

One of the outstanding advantages of single-step (Runge–Kutta) integration methods is the relative ease with which they can be made to negotiate certain types of discontinuity in a trouble-free manner. If the beginning of an integration step is made to coincide with the discontinuity, then there is no change of direction *within* the step itself. Since a single-step method requires no memory of values from past steps, no apparent truncation error is sensed by the algorithm, and there is no resulting step size reduction.

7.2.2 Multiple Step Methods

There is no easy way to overcome the effects of a discontinuity on a multiple step integration routine, other than by using one that allows for variation in the number of past steps taken into account in the calculation (a *variable-order* method). If this is done, then by reverting to single-step working, it is possible to overcome discontinuity effects in the same way as with Runge–Kutta methods. The way in which this scheme is applied is as follows. The integration step size is reduced to coincide with the discontinuity, and a normal multiple step integration takes place up to that point. This does not therefore cause any real problem. It is, however, in the need to advance away from the discontinuity again that we run into trouble with multiple step methods. The memory of past steps taken into account would linger on in the routine, and continue to cause error estimation problems. It is precisely for this reason that multiple step routines such as Gear's method are so useful. By forcing such a routine to re-initialize at a discontinuity, it reverts to first order (because it retains no past values), and behaves as though it were beginning a new integration in moving away from a discontinuity.

7.2.3 Locating Discontinuities

Before we can adjust integration step length to bring about coincidence of the end of a step with a discontinuity, we must know in advance at what point in our simulation timescale that discontinuity will occur. To this end we have to recognize that some discontinuities will be able to be located in advance of their

Table 7.1

Simulation time t	Value of state variable x	Step size h	Error estimate err	Derivative estimate dx_1	Derivative estimate dx_2	Derivative estimate dx_3	Remarks
0.0	0.0	0.1	0.0	3.0	3.0	3.0	
0.1	0.3000	0.1	0.0	3.0	3.0	3.0	
0.2	0.6000	0.1	0.0	3.0	3.0	3.0	
0.3	0.9000	0.1	0.3	3.0	−1.5	3.0	
		0.0134	0.0	3.0	3.0	3.0	
0.3134	0.9403	0.0558	0.1613	3.0	−1.5	3.0	$h_{new} \leqslant 4 \times h_{old}$
		0.0089	0.0	3.0	3.0	3.0	
0.3223	0.9670	0.0356	0.1067	3.0	−1.5	3.0	$h_{new} \leqslant 4 \times h_{old}$
		0.0067	0.0	3.0	3.0	3.0	
0.3291	0.9872	0.0270	0.0810	3.0	−1.5	3.0	$h_{new} \leqslant 4 \times h_{old}$
		0.0056	0.0084	3.0	3.0	−1.5	
		0.0025	0.0	3.0	3.0	3.0	
0.3316	0.9947	0.0099	0.0298	3.0	−1.5	3.0	$h_{new} \leqslant 4 \times h_{old}$
		0.0029	0.0043	3.0	3.0	−1.5	
		0.0016	0.0	3.0	3.0	3.0	
0.3332	0.9995	0.0064	0.0191	3.0	−1.5	3.0	$h_{new} \leqslant 4 \times h_{old}$
		0.0010	0.0031	3.0	−1.5	3.0	

t		h	err				
0.3335	1.0002	0.00064	0.0019	3.0	3.0	3.0	*err* < tolerance (0.001)
0.3339	0.9997	0.000463	0.0014	3.0	3.0	3.0	Discontinuity passed
0.3353	0.9974	0.000374	0.0011	3.0	3.0	3.0	$h_{new} \leq 4 \times h_{old}$
0.3413	0.9885	0.000324	0.00049	3.0	3.0	-1.5	$h_{new} \leq 4 \times h_{old}$
0.3650	0.9529	0.00037	0.0	-1.5	-1.5	-1.5	$h_{new} \leq 4 \times h_{old}$
0.4	0.9004	0.0015	0.0	-1.5	-1.5	-1.5	To end at $t = 0.4$
0.5	0.7504	0.0059	0.0	-1.5	-1.5	-1.5	
0.6	0.6004	0.0237	0.0	-1.5	-1.5	-1.5	
0.7	0.4504	0.0350	0.0	-1.5	-1.5	-1.5	
0.8	0.3004	0.1	0.0	-1.5	-1.5	-1.5	
0.9	0.1504	0.1	0.0	-1.5	-1.5	-1.5	
1.0	0.00043	0.1	0.0	-1.5	-1.5	-1.5	
		0.1	0.0	-1.5	-1.5	-1.5	
		0.1	0.0	-1.5	3.0	-1.5	
		0.1	0.3	-1.5	3.0	-1.5	
		0.0134	0.0403	-1.5	3.0	-1.5	
		0.0035	0.0106	-1.5	3.0	-1.5	
		0.0014	0.0043	-1.5	3.0	-1.5	
		0.00080	0.0024	-1.5	-1.5	-1.5	
		0.00054	0.00080	-1.5	3.0	-1.5	
1.0005	0.000031	0.00052	0.0	3.0	3.0	3.0	*err* < tolerance (0.001)
1.0011	0.0016	0.0021	0.0	3.0	3.0	3.0	Discontinuity passed
							$h_{new} \leq 4 \times h_{old}$

occurrence, while others will not. If a discontinuity point can be predicted, then there should be no problem in step size adjustment to suit. If a discontinuity occurs without warning, then the means have to be found to stop, locate the discontinuity accurately within the current time step, backtrack to the beginning of that time step and adjust the step size for discontinuity coincidence. Methods based on so-called *discontinuity functions* have proved useful in detecting the presence of and locating discontinuities (Carver, 1978, Bennett, 1980). The simulation language ACSL differentiates between *time events* and *state events* as far as discontinuities are concerned. The former are predictable event times that may locate in advance a series of discontinuities which may then be ordered in an event queue (see Section 3.2.2). The latter are events that bring about a change of model state, which are caused by the model itself during the execution of the simulation. These latter events use a discontinuity function approach in their detection and location.

7.2.4 Sawtooth Generation – an event-based approach

Instead of relying on the sawtooth waveform model to negotiate the peak and trough discontinuities by step size reduction, we may invoke the concept of event processing to set up in advance events that predict the occurrence of peaks and troughs in the waveform, and put this to good effect in adjusting step size to effect coincidence with peaks and troughs with the ends of integration steps. Briefly, the way in which event processing may be implemented is as follows:

- At a particular peak or trough (trough at time $t = 0$), process the event that has just occurred, by changing the value of the waveform slope.
- Schedule the next event in sequence by using the new value of slope to calculate the next event, according to the relationship

$$\text{Interval to next event} = \frac{(\text{peak value} - \text{trough value})}{|\text{slope}|} \qquad (7.5)$$

This is exactly in accordance with the rules for event processing set out in Chapter 3.

An ACSL model was written to implement the event-processing approach (the files SAWTEV.CSL and SAWTEV.CMD are listed in Appendix A.1). Figure 7.8 shows a post-run strip plot for the sawtooth waveform and slope shown separately. The event processing appears to have produced a very good-quality waveform in both cases.

The equivalent MATSIM model is listed as files SAWTEVIN.M and SAWTEVB.M in Appendix A.2. The results obtained from running a simulation from this model are shown as strip plots in Figure 7.9. Once again, the output looks to be of good quality and accurate in timing. Figure 7.10 shows the sawtooth waveform as a series of points generated at every integration step.

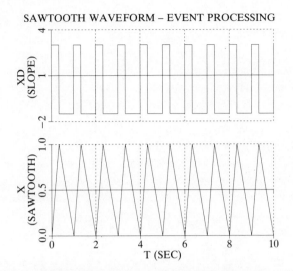

Figure 7.8 Post-run ACSL plots of event-processed sawtooth waveform and slope.

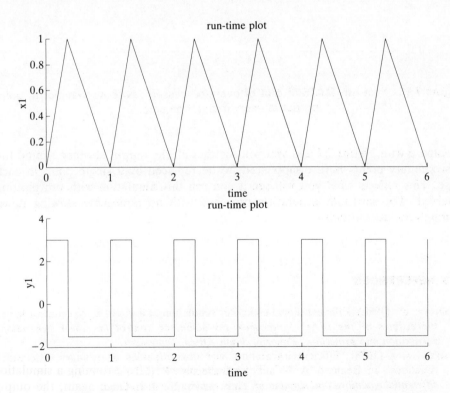

Figure 7.9 Post-run MATSIM plot of event-processed sawtooth waveform and slope.

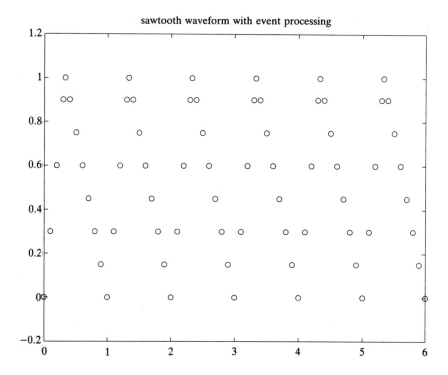

Figure 7.10   Post-run MATSIM plot of event-processed sawtooth waveform, point output at every integration step.

Compare with Figure 7.7 and you will see that all the output clusters around the discontinuity points have disappeared, to be replaced by a single point in each case. This reflects what you will see if you run this simulation with run plotting enabled. The sawtooth generation proceeds with no noticeable slowing down through the discontinuities.

7.3 REFERENCES

BENNETT, B S (1980) 'Efficient digital computer simulation of a direct digital control loop', *Proceedings of the IFAC Symposium on automatic control in power generation, distribution and protection, Pretoria, South Africa, September.*

CARVER, M B (1978) 'Efficient integration over discontinuities in ordinary differential equations', in Bennett, A W and Vichnevetsky, R (Eds) *Numerical methods for differential equations and simulation*, Amsterdam: North-Holland.

8

Digital Simulation of Sampled-Data Control Systems

8.1 INTRODUCTION

We saw in Chapter 4 how a hybrid computer makes a very good tool for the simulation of computer control systems under direct digital control. The digital part of the hybrid computer emulates the action of the control computer, the analog part replicates the process dynamics, and the interface reproduces most, if not all, of the data transfer and conversion delays inherent in the real hardware of computer control. To be successful in this respect, what digital simulation needs is a way of obtaining the same quality of replication of all these effects, solely by means of digital computer simulation techniques. There is no real problem in solving the differential and algebraic equations that constitute the process dynamics – indeed the accuracy of solution will be much better than is possible on an analog computer. The real difficulty lies in reproducing all the facets of behaviour exhibited by computer control hardware and software. The main aspects we have to take into account are:

- Replication of computer control algorithm updating at the appropriate sampling instants (zero-order hold operation)
- Execution time for control algorithm computation on the control computer
- Data conversion times for analog-to-digital (A/D) and digital-to-analog (D/A) conversion processes.

The first of these is of major concern for digital simulation. We shall see later that we may deal better with the others as refinements of our proposed method for handling the zero-order hold operation.

8.2 DISCONTINUITIES INTRODUCED BY SAMPLED-DATA SYSTEMS

In keeping with our philosophy of adhering to the simple approach to simulation problems, we can envisage direct digital control operation as follows:

- At a particular sampling instant, the A/D converter makes available to the control computer a sampled value of the controlled variable. We may assume at this stage that it has taken place with negligible A/D conversion delay.
- The control computer computes the process error with reference to a set point value (desired value for the controlled variable).
- Using the computed error, the control algorithm produces a digital value of the control action to be taken during the sample interval. This remains constant throughout the sample interval, and is made available to the process control actuator through the medium of a D/A converter. We may assume that the computation and the D/A conversion times are negligible, so that in this simple approach, error sampling and control output updating both take place at the same instant of time, the beginning of a sample interval. The control output remains constant throughout the sample interval until it is updated at the next sampling instant.

In control terms, we are seeking to implement some replication of zero-order hold operation to simulate the action of computer control. This might seem simple enough to do in theory, but practical implementation in a digital simulation causes many problems, both with the introduction of discontinuities to the system transients and with sample timing errors that may accrue through injudicious methods of timing operation.

Figure 8.1 illustrates discontinuities introduced by the zero-order hold operation. We may consider this to be the action taken by straight sampling and process actuation according to the value of the process error (i.e. proportional control with unity gain). The fact is that whatever the control algorithm involved, the zero-order hold nature of the control output introduces discontinuities into the process. We saw in the previous chapter that discontinuity effects are worst when they are directly involved in the evaluation of derivatives for the model differential equations. A common example of this occurrence in process control is the case when the digital compensator output forms the discontinuous derivative input to a process control valve actuator, as shown in Figure 8.2. Figure 8.2(a) shows the overall block diagram for a very simple sampled-data control system. Figure 8.2(b) shows the same diagram with the valve block replaced by a simple transfer function that depicts the valve as an integrator with gain, surrounded by its position feedback. It is clear, therefore, that there are advantages for us in resorting to the methods outlined in the previous chapter. We may use them to alleviate discontinuity effects of sampled-data control upon the speed of numerical integration. This is especially true when sampling periods are short, that is sampling frequencies are high.

In process control, it is fortunate that sampling instants are for the most part entirely predictable in advance. Sampling may commonly be single rate, multirate or cyclic. Whatever form we use, the sampling mechanism is set in motion by the control computer. This indicates that it can be planned ahead, that is the sampling

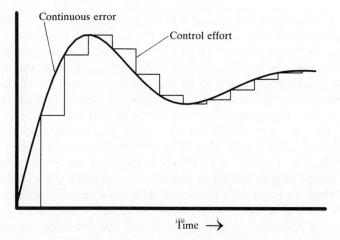

Figure 8.1 Discontinuities caused by zero-order hold operation.

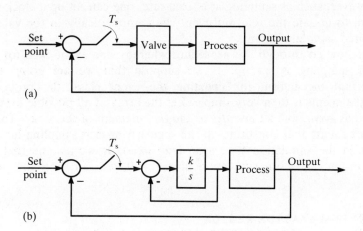

Figure 8.2 (a) Block diagram of simple sampled-data control system; (b) valve block replaced by integrator with feedback.

instants are programmed before their occurrence. It should therefore be possible to set up an event queue of sampling instants for the whole time span of a simulation run.

8.3 EVENT PROCESSING FOR SAMPLED-DATA SYSTEM SIMULATION

Let us consider the simplest type of sampled-data operation first, namely single-rate sampling with a constant interval between samples. Given that the interval size is predetermined, one could pre-program all sample times for the whole simulation run. This would, however, require a large amount of memory to store a look-up table involved with high-frequency sampling. Fortunately there is an easy way around the problem that is borrowed again from the concepts of discrete-event simulation. We set up an array of event times for all the different classes of events that may occur during a continuous system simulation. We may set the size of the array simply by inclusion of the algorithm that defines each event class. This is very similar to the event processing that is the main mechanism behind event-based discrete-event simulation languages like GASP (Pritsker, 1974). The array is then simply maintained on a 'next-event' basis, whereby the stored entries constitute the next events to be processed in each of the classes. When an event of a particular class is processed, the corresponding element of the array is updated to the value of the next event time for that class. Furthermore, this updating can take place in two possible ways. It may be by a user-defined algorithm, which is useful if the interevent time is a function of other parts of the simulation and changes value accordingly. For constant interevent times, however, such as sampling at a fixed rate, one can set up a clock-interrupt mechanism to update the relevant event class automatically to the value of the next required sampling instant.

Let us now go through the procedure whereby one event class for constant single-rate sampling is set up. If we suppose that we are going to use a clock-interrupt mechanism for updating the event class, the procedure for handling the events is then very simple. Let the array of all possible event classes be labelled as *event*, and let *event[i]* be the $(i)^{th}$ element of the array. This will be the class of events that constitute all the regularly spaced sampling instants that will occur in the simulation time span (*run time*). If we assume that the first sampling instant occurs at simulation time $T = 0$, we can show the process as follows:

```
// INITIALIZATION REGION
n = 0     // Initialize the sample count and the
event[i] = 0.0 // event class
  "     "    "
  "     "    "
  "     "    "
// EVENT-PROCESSING REGION
n = n + 1 // Increment the sample counter
event[i] = n*sample_interval // Compute the next
                             // event
                             // in this class
```

If n is an integer variable, computation of the next-event time in this way will ensure maximum accuracy. The common practice of incrementing the event time by addition of the sample interval can cause the accumulation of round-off errors and a subsequent degradation in timing accuracy for the sampling process.

In the above 'coding statements' the event time updating takes place in a special event-processing region appropriate to the class representing the sampling instants. It now remains to determine what form the actual event processing takes, and what its effect is likely to be on the behaviour of the simulation. In the previous chapter, in dealing with the example of the sawtooth generator, we saw how the square wave representing the derivative of the sawtooth function was 'aliased' within conventional graphical output for the two waveforms. We also saw how the use of event processing removed the aliasing and achieved the ability to create a square wave that was truly double valued at the event times when the derivative value altered. The same result occurs with the sampling process, when we include within the event-processing region the control algorithm for sampled-data compensation. The discontinuous changes in digital compensator output that give a zero-order hold characteristic will also be truly double valued at the sampling instants, and yield a staircase type of display similar to that shown in Figure 8.1. We achieve the double-valued nature of the output by arranging for the transient values to be sent to the output device (screen and/or disk file) before the event is processed, and again after processing is complete. For a control algorithm of the form

$$u_n = a_1 u_{n-1} - a_2 u_{n-2} + b_0 e_n - b_1 e_{n-1} + b_2 e_{n-2} \tag{8.1}$$

we may code an event-processing region as follows:

```
err[2] = err[1]   // Update shift registers for
                  // sampled error and compensator
err[1] = err[0]   // output to move past values down
                  // by one place. In this listing
com[2] = com[1]   // err[0-2] = e(n) to e(n-2)
                  // and com[0-2] = u(n) to u(n-2).
com[1] = com[0]   //
err[0] = sample   // Sample the error and update the
                  // compensator output
com[0] = a1*com[1] - a2*com[2] + b0*err[0] -
         b1*err[1] + b2*err[2]
```

The compensator output remains constant until the next sampling instant when the event processing is repeated. Figure 8.3 shows the output sequence when an event of this nature is processed.

It is important to realize that the output sequence that takes place at an event time is quite independent of that which occurs at the designated communication intervals. We have already seen how the use of an event-processing approach gave us the ability to tailor our integration step lengths so that an output point coincides with the end of an integration step. We can see, therefore, that what we are doing at a communication point is to sample the (supposedly continuous) model transients and output the sampled values to the output device in use at the time. The imposition of sampled-data control action is really another layer of the

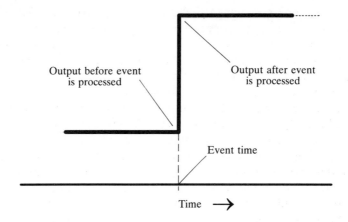

Figure 8.3   Portion of zero-order hold transient showing event processing.

same approach to the treatment of integration step length. The only real difference for dedicated event processing is that we sample the model transients twice at each event, once before the event and once afterwards.

8.4 A SIMPLE SAMPLED-DATA CONTROL SYSTEM

Figure 8.4 shows a system similar to that in Figure 5.2 with a sampler inserted in the forward path before the process. This is a single-rate sampled-data control system, whereby the control is obtained simply by closing the loop. The process may be treated in the same manner for the purposes of simulation, decomposing into two separate first-order blocks. An ACSL model for this process is listed in Appendix A.1 (files SAMPLE.CSL and SAMPLE.CMD).

Figure 8.5 gives the results of a run-time plot from an ACSL run with this model. The sampling period is 0.8 seconds, and with this in mind, it is fairly easy to check the transients for correctness. For example, X_1 is the integral of the sampled error multiplied by 10, so that the value at the end of the first sampling interval should be 8.0 for a sampled error value of 1.0. This error is derived from sampling the continuous error every 0.8 seconds, and, at time $t = 0$, the continuous error is equal to the set point value of 1.0 (since $X_2 = 0$ at time $t = 0$). Similar checks may be made at other points during the simulation timescale. Also, the zero-order hold action of the sampler appears to be correct, in that the staircase function is truly double valued at the sampling instants (owing to implementation of the LOGD facility in ACSL). The part of the simulation that is most difficult to check from the transients in Figure 8.5 is the relationship between X_1 and X_2. However, in simulation done properly, other means of

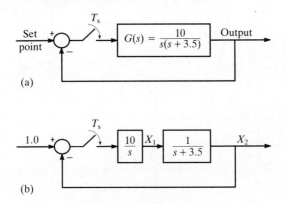

(a)

(b)

Figure 8.4   A simple sampled-data control system.

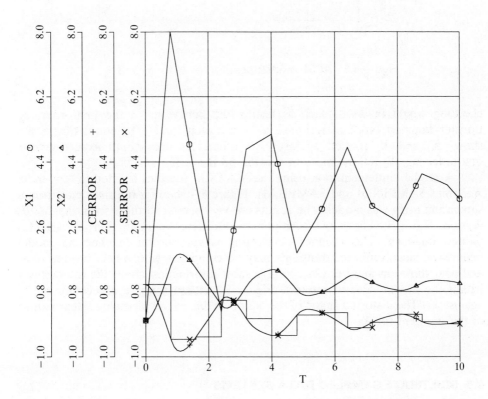

Figure 8.5   ACSL model: run-time plot for sampled-data system.

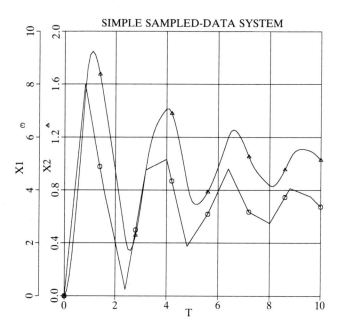

Figure 8.6   ACSL model: post-run plots for x_1 and x_2.

checking would be used, such as testing the behaviour of the first-order lag transfer function block in response to a unit step forcing function. Figure 8.6 shows X_1 and X_2 plotted against each other. As one would expect with a first-order lag, X_2 follows more or less the profile of X_1.

A model implementation using the MATSIM routines is listed in Appendix A.2 (files SAMPIN.M and SAMPB.M). Figure 8.7 shows a run-time plot from a simulation using this model. The results are very similar to those from the ACSL version. Figure 8.8 shows two strip plots with different combinations of variables plotted on each. This manner of plotting is very useful in checking model behaviour, since different transients may be compared with others on one plot, and with those on another plot, to give most information about the interactions within the model. Out of interest, Figure 8.9 shows a phase plane plot of X_2 against X_1. The distorted shape of the 'spiral' is due to the piecewise linear nature of the transient for X_1.

8.5 MULTIRATE SAMPLED-DATA SYSTEMS

On the foundations that we have laid for single-rate sampling, it is easy enough to build an event-processing structure that caters for multirate sampled-data control.

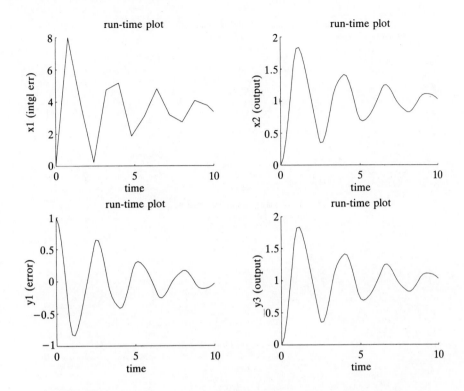

Figure 8.7   MATSIM model: run-time plot for sampled-data system.

All that we need is a separate event class for each sampling rate present in the system. A typical multirate sampled-data control system will have the error sampled at one rate and the compensator output updated at a different rate, usually an integer fraction or multiple of the first. If the two sample rates are constant, the event processing can be mechanized by means of a clock interrupt in each case. The sample rates may or may not be synchronized, but it is easier at first to consider them to operate synchronously. Consider the block diagram of Figure 8.10. This figure depicts a multirate sampled-data system (Kuo, 1963a) in which two samplers operate within a single-input–single-output control system.

The sampler on the compensator output operates N times faster than the error sampler. What this means qualitatively is that the digital compensator output will tend to respond to the error that is sampled far more quickly than it would if its output were updated only at the error sampling rate. The rapid updating of the compensator output constitutes what is in effect an iterative loop. This attempts to reach an equilibrium value of control effort corresponding to the error sample that is taken at the slower rate. We can see the effect of this in the following example: for a simple first-order lag, $D(s) = 1/(s+1)$. In a continuous

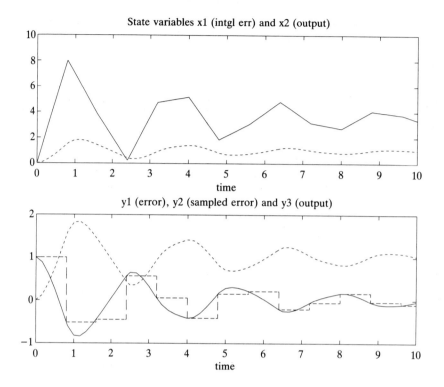

Figure 8.8   MATSIM model: various sampled-data system responses.

sense, we would expect the output of $D(s)$ to approach exponentially the error value that is sampled every period of T seconds. Since the corresponding digital compensation algorithm is given by the expression

$$u_n = 0.368u_{n-1} + 0.632e_{m-1} \qquad\qquad (8.2)$$

for a sample period $T = 1$ second, we can see the effect digitally of the same exponential approach from the values given in Table 8.1. For the purpose of computing the entries in this table, we let the sample ratio $N = 3$.

Each time the error is sampled, the compensator output undergoes a rapid iteration to converge towards that sampled error value. Although Table 8.1 makes it appear that the count on the rapid sampling is reset every error sample cycle, this is not the case under event-processing control. We need to set up two separate event classes, one faster than the other by a factor N (the sample ratio). The control effort is continuously updated, using as input whatever value of the error is current at the time. The two event-processing cycles operate independently of each other, and may be synchronous or asynchronous as desired.

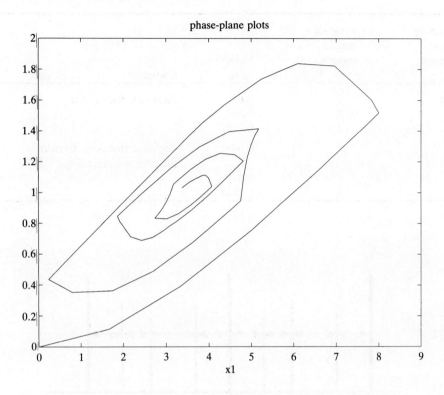

Figure 8.9   MATSIM model: phase plane plot of x_2 against x_1.

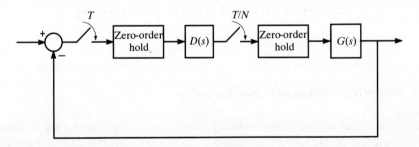

Figure 8.10   Block diagram of multirate sampled-data control system.

Table 8.1   Output values for the example of a multirate sampled-data system

Value of error sample e_{m-1}	Compensator output sample n	Value of u_n	Comments
1.0	1	0.632	Assumes that $u_0 = 0$
	2	0.865	
	3	0.950	
	1	0.984	Assumes that $u_0 = u_3$ from
0.5			the previous iteration
	2	0.678	
	3	0.5655	

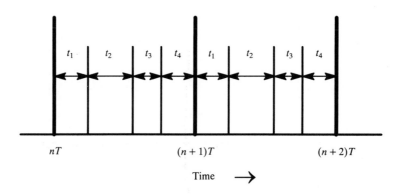

Figure 8.11   Timing diagram for cyclic sampling.

8.5.1 Digital Simulation of Cyclic Sampling

A particular form of multirate sampling exists where the faster rate is variable within one cycle of the slower sampling rate. When the faster rate is periodic in nature, that is when the pattern of variability is repeated every slow sampling period, this is called cyclic sampling (Kuo, 1963b), an example of which is shown in Figure 8.11.

It is not difficult to set up an event-processing scheme whereby the slower sampling operates under clock interrupt, while the variable faster sampling carries on under event-processing control. We have to set up an array to contain a list of all the variable sample intervals in one slow sample interval. Every time a slow-rate sample is taken, the list is initialized to start again from the beginning. The following example, based on the timing diagram of Figure 8.5, shows how the events are processed in this case:

```
// Event-processing region for slow (error)
// sampling - to set event[i]
    m = m + 1            // Increment sample counter
                        // for fixed-rate sampling
    evthld = event[i]   // Store the current value
                        // of time (the event time)
    event[i] = m*T      // Compute time of next event
                        // in this class
    n = 0               // Reset counter for cyclic sampling
    err = sample        // Sample the continuous error

// Event-processing region for cyclic sampling
    n = n + 1           // Increment cyclic sampling counter
    if (n <> 4)         // Set next event in the cycle, by
                        // adding on
    event[j] = event[j] + t[n]// variable cyclic
                        // sample periods, except
    else                // in the case of the last
                        // such period, set equal
    event[j] = event[i] // to the next error
                        // sampling event.
    u[n] = expression   // Update compensator
                        // output
```

In the above program, we should note the following points. In any multirate sampling operation, we have to take care to ensure that the event processing takes place in the correct order. In the example, we have assumed that error sampling precedes the initiation of the cyclic sampling iterative loop, and hence that event-processing region takes precedence in the coding sequence. Also we should remember that we have assumed the cyclic sampling initiation to be synchronous with the error sampling. This is the reason for setting the last cyclic sampling event equal to the next error sampling time, so as not to build up any timing errors. In the above program, T is the slow sampling period and $t[1-4]$ are the cyclic sampling periods.

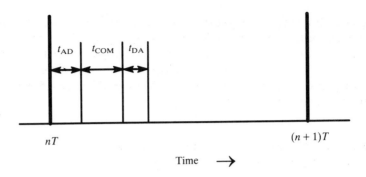

Figure 8.12   Sampled-data system timing diagram of control hardware data conversion and processing delays.

8.6 APPLICATION TO COMPUTER CONTROL PROBLEMS

So far in dealing with the question of sampled-data control, we have made the assumption that any delays caused by computation and data conversion are negligible. As we were aware in our dealings with hybrid computation, the real hardware and software of sampled-data control do lead to these delays. In many cases with fast-acting control action, where we try to emulate a real sampled-data control system, we do need to try to include these effects in our digital simulation representation of the control action.

Let us again consider a single-rate sampled-data control system within which we will incorporate delays due to algorithm execution and data conversion times. Let us assume that the sequence of events is similar to that given at the beginning of Section 8.2, but this time with the computation and data conversion delays being taken into account. Figure 8.12 shows a typical timing diagram that we might postulate for this operation. In this diagram, T is the error sampling period, t_{AD} is the A/D conversion time, t_{COM} is the total computation time and t_{DA} is the D/A conversion time. The total computation time consists of the time taken to compute the error (since it is usually the controlled variable that is sampled) and the compensator algorithm execution time. For display purposes, these delay times look very exaggerated. In reality they will be very small indeed compared with the error sampling period, for most applications. If the sampling is very fast, however, then these delays may become more important and lead to a significant overall time delay in between the sampling of the controlled variable and the updating of the control effort.

If we are to take these delays into account, it is usually quite in order to assume that they are constant in value. This means that we may add all three values together to yield a single delay time, t_{TOT}, that we can use to set an event for updating the control effort according to the scheme below.

```
// Event-processing region for error sampling
// (under clock-interrupt control)
e[n] = sample    // Sample the error
n = n + 1        // Increment the sample counter
event[j] = event[i] + t_tot // Set the next
                            // event for control
                            // effort update
event[i] = n*T   // Compute the next event in
                 // this class

// Event-processing region for compensator output
// (control effort) updating

u[n] = expression // Update the control effort
```

We can see from this scheme that the first event-processing region sets not only the next event for error sampling, but also that for control update. The second such region does not itself set any next event times.

8.7 EXAMPLE: MULTIRATE SAMPLED-DATA SYSTEM

Figure 8.13(a) shows a block diagram for a continuous single-input–single-output control system with unity feedback. The process consists of a second-order process actuated by a control valve (with rate limiting, but without position feedback). A double phase advance compensator completes the forward path for the loop. In this example we have the following relationships:

Process transfer function

$$G_p(s) = \frac{1}{s^2 + 4s + 8} \tag{8.3}$$

Compensator transfer function

$$\frac{U(s)}{E(s)} = G_C(s) = K\frac{(1 + T_1 s)(1 + T_2 s)}{(1 + T_3 s)(1 + T_4 s)} \tag{8.4}$$

where the compensator time constants are given as

$T_1 = 0.5, \quad T_2 = 0.5, \quad T_3 = 0.1, \quad T_4 = 0.1$

$U(s)$ is the Laplace transform of the control output
$E(s)$ is the Laplace transform of the measured temperature error
K is the control gain ($K = 2.0$)

If we define a sampling period T_s, it may be shown that the following digital controller algorithm is equivalent to the continuous compensator given in equation (8.4) above. This is a compensator similar to that used in the case study presented in Chapter 10.

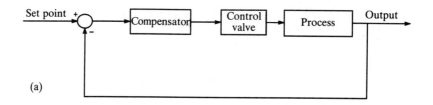

(a)

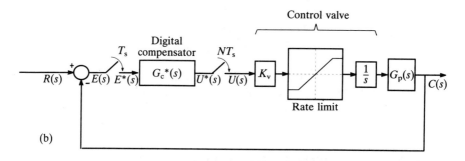

(b)

Figure 8.13   (a) Block diagram of unity-feedback SISO control system; (b) block diagram expanded into multirate sampled-data system with rate-limited control valve.

Digital controller algorithm

$$U_N = K\frac{A_2}{A_1}\left(E_N - \frac{B_2}{A_2}E_{N-1} + \frac{C_2}{A_2}E_{N-2}\right) + \frac{B_1}{A_1}U_{N-1} - \frac{C_1}{A_1}U_{N-2} \tag{8.5}$$

where

$$A_1 = \frac{(T_s + T_3)\cdot(T_s + T_4)}{T_s^2} \tag{8.6}$$

$$A_2 = \frac{(T_s + T_1)\cdot(T_s + T_2)}{T_s^2} \tag{8.7}$$

$$B_1 = \frac{(T_3 + T_4)T_s + 2T_3T_4}{T_s^2} \tag{8.8}$$

$$B_2 = \frac{(T_1 + T_2)T_s + 2T_1T_2}{T_s^2} \tag{8.9}$$

$$C_1 = \frac{T_3T_4}{T_s^2} \tag{8.10}$$

$$C_2 = \frac{T_1T_2}{T_s^2} \tag{8.11}$$

and

U_N is controller output at present time
U_{N-1} is controller output at last sampling time
U_{N-2} is controller output at penultimate sampling time
E_N is temperature error at present time
E_{N-1} is temperature error at last sampling time
E_{N-2} is temperature error at penultimate sampling time

Figure 8.13(b) shows a block diagram of the system with all variables shown in Laplace notation, amended to include a breakdown of the valve characteristics. This shows that the compensator output after sampling represents a valve motor rate demand. Since there is a limit on the rate at which the valve motor can operate, a non-linear block is included to represent that rate limit. Finally the limited rate is input to an integrator to generate valve position resulting from motor actuation.

The implementation of multirate sampled-data compensation now needs to be explained. In Section 8.4, we saw how a faster sampling of the compensator output can result in quicker updating of the effect of the sampled error. This example makes use of entirely different reasoning to justify sampling the error at a rate faster than the compensator output.

In equation (8.5) we can see how information is required that is out of date by two sampling periods. If we make those sampling periods shorter, then the information becomes more up to date and better control should result. It is expedient, therefore, to implement the compensation in the following manner:

- Select the sampling rate at which compensator output to the control valve is to be updated.
- Choose a faster sampling rate at which the continuous error is to be sampled (an integer multiple of the compensator sampling rate is expedient).
- Update the compensator output at this faster sampling rate.

If the sampling rate is N and the faster sampling period for error sampling and compensator update is T_s, then the compensator output will be sampled every NT_s time units. For this example, $N = 3$ and $T_s = 0.2$ time units.

An ACSL model listing is given in Appendix A.1 (files MULTR.CSL and MULTR.CMD). Figure 8.14 shows a post-run ACSL plot of some transient responses to a unit step change in set point. The output (solid line) is reasonably well behaved and is shown with dotted-line responses for the continuous and sampled errors. The error sampling period can be seen to have the correct value $T_s = 0.2$ time units. Figure 8.15 shows the transients that represent valve operation. The compensator output (correctly sampled every 0.6 time units) is, in effect, the demanded valve motor rate (a value of about 22 in the first sampling period). The rate-limited value is correctly shown as 15, so this shows that the limiter is working correctly. Elsewhere in the simulation run, the demanded valve

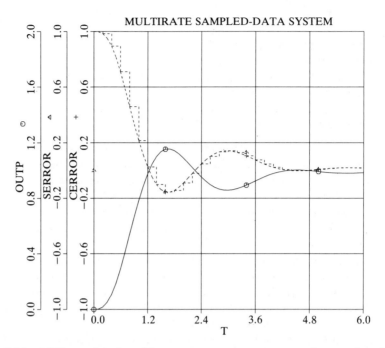

Figure 8.14   ACSL model of multirate system: post-run pots of several variables.

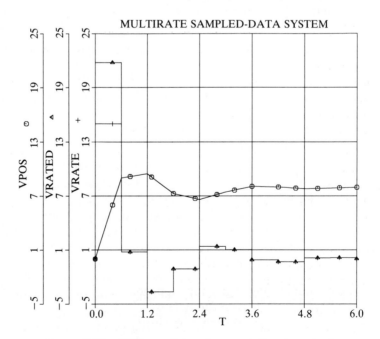

Figure 8.15   ACSL model: valve operation transients.

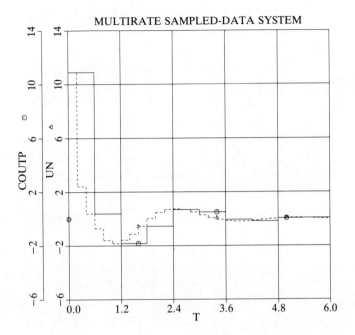

Figure 8.16   ACSL model: outputs of the two samplers.

motor rate is less than 15. The valve position is correct in terms of the slope of its piecewise linear movements every sampling period. Lastly, but not least, Figure 8.16 illustrates the relationship between the fast sampling rate at which error sampling and compensator output generation take place, and the slow sampling rate at which that compensator output is sampled as a valve rate demand. We can see that the relationship is working correctly.

If you, the reader, feel that you have been deluged with transient responses from this example, this is because, in the development stages of a model at least, it is important to have as much output information as possible to check every aspect of the model behaviour against expectations. This constitutes the *model verification* that was discussed in Chapter 2. We now know that the model behaves as we expect it to behave from programming considerations. What we do not know is whether or not it is an adequate representation of the real physical system. The validation required to show this is outside the scope of this book.

A MATSIM model listing is given in Appendix A.2 (files MULTRIN.M and MULTRB.M). Figure 8.17 shows a run-time plot for four of the system variables. The model behaviour is identical to that obtained from ACSL.

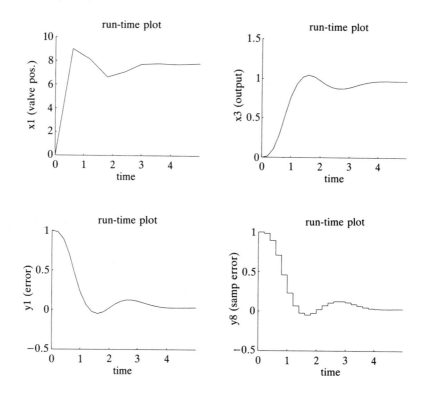

Figure 8.17   MATSIM model of multirate system: run-time plot of several variables.

8.8 EXAMPLE: MULTIRATE SAMPLED-DATA SYSTEM WITH NOISE

As an extension to the above example for a multirate sampled-data system, event processing that was so successfully used to mechanize the sampling operations can be used to good effect to generate a noise function to be added to one of the simulation variables. If we consider that we wish to simulate the effect of having a noisy measurement of the output variable, then we may represent this by adding a noise signal to the continuous error which we can then sample in the normal way for the purpose of compensator updating.

If we want to make it easy for ourselves, we will choose a noise function that is generated by an algorithm that is updated at regular intervals, thereby introducing an extra level of sampling. This sampling will have to be at a sufficiently high rate, in order to represent noise that covers an adequate range of frequencies. We should choose a noise sampling frequency that is somewhat higher than the highest frequency that we expect to find in our model.

Let us choose as the random variable in our noise generation the amplitude

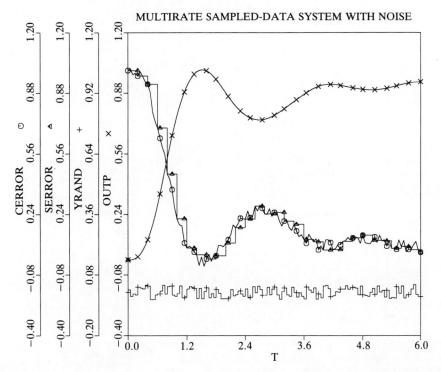

Figure 8.18 ACSL model of multirate system with noise: post-run plots of several variables.

that it will have within a noise sampling period. If we draw a sample from a uniform distribution (see Chapter 3) we may use that sample to determine the amplitude of the noise signal over the next noise sampling period. A random-number generator that generates values between 0 and 1 will give noise that has a mean amplitude of 0.5 and a standard deviation of $\sqrt{1/12}$. Giving the noise a zero mean and a standard deviation D requires application of the formula

$$N_{\text{required}} = (N_{\text{standard}} - 0.5) \times D\sqrt{12} \tag{8.12}$$

Implementation of this noise generation within the multirate sampled-data control system covered in the previous section requires an extra event-processing capability to be added to the model. In this example, we use a standard deviation $D = 0.2$ and a noise sampling period $T_n = 0.05$ time units.

An ACSL model is listed in Appendix A.1 (files MULTN.CSL and MULTN.CMD). Figure 8.18 shows transient responses to a unit step change in set point. The output is a little more oscillatory than its counterpart in Figure 8.14, but otherwise the responses are similar. The noise characteristic is shown at the bottom of the figure. The variable amplitude and regular switching are quite evident.

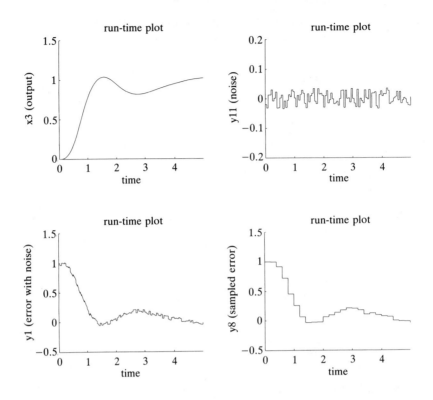

Figure 8.19   MATSIM model of multirate system with noise: run-time plots of several variables.

An equivalent MATSIM model is listed in Appendix A.2 (files MULTNIN.M and MULTNB.M). Figure 8.19 illustrates the run-time plots obtained for the four variables that feature in Figure 8.18. Notice the differences in the shape of the output response. This is to do with the noise characteristic that was superimposed upon the continuous error. This is different (because it started from a different seed value) and it therefore had some bearing on the error values that were sampled. This illustrates a point about the inclusion of stochastic effects in continuous simulations. In order to achieve an average pattern of behaviour, either we have to run a simulation for long enough as we would for discrete system models, or we would have to repeat a short run many times with different seed conditions to obtain many realizations of the system response.

Finally, as part of this example, we can use the facilities present in MATLAB to obtain a power spectral density plot. An example of this for the MATSIM model is shown in Figure 8.20. This shows that the noise generated is not exactly

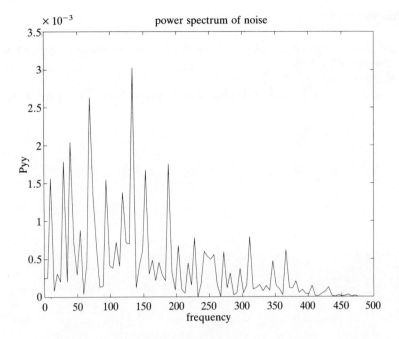

Figure 8.20   MATSIM model: power spectral density of noise.

'white' in that there are several 'peaks' that predominate. Running for longer, with perhaps a higher sampling frequency, could produce a better-quality band-limited noise.

8.9 CONCLUSIONS

In this chapter we have seen that the concepts of event processing borrowed from discrete-event simulation can prove very useful in the digital simulation of sampled-data systems of all kinds. This is from the point of view of both negotiating the discontinuities involved, and the benefits that accrue from this approach, namely potentially greater accuracy and anti-aliasing of the output of simulation transients. We should remember, however, that the refinements we can apply to take into account all the hardware and software delays of computer control should be looked at with some degree of scepticism in all but the fastest control systems. Even in these, the benefits of adding these effects to a simulation may be offset by the inaccuracy with which the delay data are known. One would really have to carry out hardware and software tests to establish the delay times with any degree of precision, should one wish to take them into account.

8.10 REFERENCES

KUO, B C (1963a) *Analysis and synthesis of sampled-data control systems*, Englewood Cliffs, NJ: Prentice Hall, p. 368.

KUO, B C (1963b) Ibid., p. 370.

PRITSKER, A A B (1974) *The GASP IV simulation language*, New York: John Wiley.

9

Distributed-Parameter Systems

9.1 PARTIAL DIFFERENTIAL EQUATION MODELS

Many dynamical systems are represented for the purpose of simulation by distributed-parameter models, being both time dependent and space dependent in terms of one or more distance variables. In this chapter, we will consider some basic types of distributed-parameter systems, and the partial differential equations that constitute the system models. To simplify matters, we shall only consider one dimension in space and the partial differential equations that arise therefrom. Partial differential equations (PDEs) often look more daunting than they really are, and for those of you new to distributed-parameter systems, a very comprehensive list of fundamental equations used in science and engineering is presented in the book by Hughes and Gaylord (1964). Most of the equations presented are PDEs. This shows that science and engineering systems, in the main, are more truly represented as distributed-parameter systems, rather than the lumped-parameter approximations that are often used. Himmelblau and Bischoff (1968) show how many of the PDEs used in process engineering may be derived from first principles. Some understanding of this is fundamental to the work of this chapter, so a similar approach is used to introduce the subject of distributed-parameter systems and PDEs in general.

In common with all the other systems with which we have dealt so far, we have to make use of some method of converting the PDEs into ordinary differential equations (ODEs), if we are to be able to perform distributed-parameter simulations using the general-purpose means at our disposal. In order to do this, we need to discretize with respect to one of the independent variables, which requires finite-difference approximations for some of the derivatives.

9.2 DISCRETIZATION: CONVERSION OF PDEs TO ODEs

In their classic book on hybrid computation, Bekey and Karplus (1968) discussed various ways in which discretization might be carried out. Their concern was for the implementation of distributed-parameter system simulations on hybrid computers, and how one or more independent variables could be discretized to allow solution on the analog portion of the hybrid computer. Schuchmann (1970) presented a very comprehensive survey of the distributed-parameter simulation scene in which he put forward well-known examples of different kinds of PDEs in one spatial dimension, together with various methods of solution.

In keeping with our requirements that modelling should result in analog-compatible sets of differential equations, we may look at solution methods that result in the reduction of PDEs to sets of ODEs, of any order. Once we have a set of ODEs, decomposition of high-order equations to first order is carried out in the normal way. Since we are dealing with time-varying distributed-parameter models of one spatial dimension only, we are left with two alternatives:

- to discretize the model in the spatial dimension, and solve the resulting set of ODEs with respect to time, or
- to discretize with respect to time, and solve the resulting differential equations with respect to the spatial dimension by means of our general-purpose simulation facilities. In this case, we need to transform simulation time into distance as the independent variable.

Let us compare these two methods by looking at simple examples, and see how they differ in ease of implementation. We may begin by examining the derivation of one of the fundamental equations of fluid mechanics – the so-called *continuity equation* for a compressible fluid. This is nothing more than an application of the Law of Conservation of Mass for the fluid, derived here for a one-dimensional flow pattern. As is often the case, it is useful to derive such an equation from basic principles as follows.

Figure 9.1 shows an element of volume in the flow path, shown for simplicity with a constant circular cross-section of area A. If the length of this element is Δx, then the volume of the element is $A . \Delta x$. If ρ is the fluid density within the element, the mass of fluid is $\rho . A . \Delta x$, and the accumulation of fluid mass within the element over a period of time Δt is given as

$$\text{Accumulation} = \rho A \Delta x|_{t+\Delta t} - \rho A \Delta x|_t \tag{9.1}$$

The input flow during this period of time Δt is given by

$$\text{Input} = \rho A V \Delta t|_x \tag{9.2}$$

where V is the linear flow velocity of the fluid, while the output flow during the same period is

$$\text{Output} = \rho A V \Delta t|_{x+\Delta x} \tag{9.3}$$

Since accumulation = input − output, we may write

$$A[\rho_{t+\Delta t} - \rho_t]\Delta x = A[(\rho V)_x - (\rho V)_{x+\Delta x}]\Delta t \tag{9.4}$$

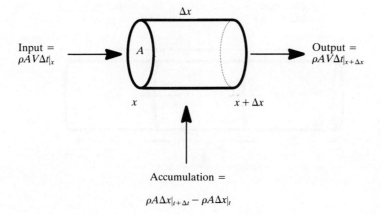

Figure 9.1 Finite element of volume for derivation of continuity equation.

We may re-arrange equation (9.4) to give

$$A\left[\frac{\rho_{t+\Delta t} - \rho_t}{\Delta t}\right] = A\left[\frac{(\rho V)_x - (\rho V)_{x+\Delta x}}{\Delta x}\right] \tag{9.5}$$

from which we obtain in the limit as $\Delta t \to 0$ and $\Delta x \to 0$

$$\frac{\partial \rho}{\partial t} = -\frac{\partial(\rho V)}{\partial x} \tag{9.6}$$

Equation (9.6) is sometimes written in a different form

$$A\frac{\partial \rho}{\partial t} = -\frac{\partial M}{\partial x} \tag{9.7}$$

where the mass flow of fluid is given by

$$M = \rho A V \tag{9.8}$$

Equations (9.6) and (9.7) are examples of PDEs that are first order in both t and x. Let us now consider the derivation of a PDE that is of higher order.

Figure 9.2 shows three finite elements along the length of a core of material (assumed solid to eliminate all heat dissipation effects other than by conduction alone). We are interested in determining the effect on conduction of heat along the length of the bar purely in an axial direction, brought about by differentials in temperature.

Let us consider the middle element whose temperature is θ_x and derive expressions for the heat input, heat output and heat accumulation within that

Heat input =
$kA\Delta t(\theta_{x-\Delta x} - \theta_x)/\Delta x$

Heat output =
$kA\Delta t(\theta_x - \theta_{x+\Delta x})/\Delta x$

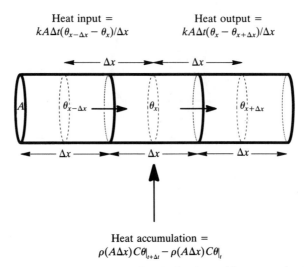

Heat accumulation =
$\rho(A\Delta x)C\theta|_{t+\Delta t} - \rho(A\Delta x)C\theta|_t$

Figure 9.2 Finite-element scheme for derivation of heat conduction equation.

particular element. The heat input during a time interval Δt, across the boundary from the element on the left of the diagram, is defined as

Heat input = (thermal conductivity $\times$ area $\times$ temperature gradient) Δt

$$= kA\Delta t(\theta_{x-\Delta x} - \theta_x)/\Delta x \qquad (9.9)$$

where k is thermal conductivity, A is cross-sectional area of the core of material, θ is temperature and Δx is the centre-to-centre distance between the elements (assumed the same as the length of an element itself). Similarly, the heat output during the time interval Δt from the central element is given by

$$\text{Heat output} = kA\Delta t(\theta_x - \theta_{x+\Delta x})/\Delta x \qquad (9.10)$$

The heat accumulation within the central element over the time interval Δt is defined as

Heat accumulation = (mass $\times$ specific heat $\times$ temperature) at time $t + \Delta t$
$\qquad\qquad$ − (mass $\times$ specific heat $\times$ temperature) at time t

This becomes

$$\text{Heat accumulation} = \rho(A.\Delta x)C.\theta|_{t+\Delta t} - \rho(A.\Delta x)C.\theta|_t \qquad (9.11)$$

where ρ is density and C is specific heat (both assumed constant).
   Now

Heat accumulation = heat input − heat output

$$\therefore \rho(A.\Delta x)C(\theta_{t+\Delta t} - \theta_t) = kA\Delta t\left[\frac{\theta_{x-\Delta x} - \theta_x}{\Delta x} - \frac{\theta_x - \theta_{x+\Delta x}}{\Delta x}\right] \qquad (9.12)$$

$$\therefore \rho C \left[\frac{\theta_{t+\Delta t} - \theta_t}{\Delta t} \right] = \frac{k}{\Delta x} \left[\frac{\theta_{x-\Delta x} - \theta_x}{\Delta x} - \frac{\theta_x - \theta_{x+\Delta x}}{\Delta x} \right] \qquad (9.13)$$

In the limit as $\Delta t \to 0$ and $\Delta x \to 0$, we obtain the PDE

$$\frac{\partial \theta}{\partial t} = \frac{k}{\rho C} \frac{\partial}{\partial x} \left(\frac{\partial \theta}{\partial x} \right)$$

$$= \frac{k}{\rho C} \frac{\partial^2 \theta}{\partial x^2} \qquad (9.14)$$

This is second order with respect to x, which we may have guessed from inspection of Figure 9.2, since we encompassed three spatial finite elements to derive equation (9.14).

We are now in a position to examine the two discretization schemes mentioned near the beginning of Section 9.2. If we look first of all at the continuity equation

$$\frac{\partial \rho}{\partial t} = - \frac{\partial (\rho V)}{\partial x} \qquad (9.6)$$

we may discretize with respect to either space or time, with reference to equation (9.5) which contains the discrete form of both. If we discretize with respect to space, we have the ordinary differential equation

$$\frac{d\rho}{dt} = \frac{(\rho V)_x - (\rho V)_{x+\Delta x}}{\Delta x} \qquad (9.15)$$

So far, we have stipulated nothing about which value of ρ to use on the left-hand side of equation (9.15). In Figure 9.1, it seems from the position of the upward arrow that we have chosen a value at the midpoint of the element of length Δx. If this were indeed the case, we would end up with a *central difference approximation*

$$\frac{d}{dt} \left(\frac{\rho_x + \rho_{x+\Delta x}}{2} \right) = \frac{(\rho V)_x - (\rho V)_{x+\Delta x}}{\Delta x} \qquad (9.16)$$

On the other hand, we may choose to go for a *backward difference solution* in which case we would have the equation

$$\frac{d\rho_{x+\Delta x}}{dt} = \frac{(\rho V)_x - (\rho V)_{x+\Delta x}}{\Delta x} \qquad (9.17)$$

The backward difference equation (9.17) is probably easier (and safer) to implement than the central difference solution, for two reasons:

- It is negative feedback with respect to the solution variable $\rho_{x+\Delta x}$ (ρ_x is an input to the finite element), whereas in equation (9.16), the equation is a mixture of positive and negative feedback. Equation (9.17) is therefore completely stable, while the stability of equation (9.16) is questionable.

- Equation (9.16) requires the derivative of the input variable ρ_x, as well as the variable itself. This may be available from a preceding element, but not necessarily at the first element in the series that make up the model.

Let us now discretize equation (9.14) with respect to space, using as reference the finite-difference equation (9.13). We then have

$$\frac{d\theta}{dt} = \frac{k}{\rho C \Delta x} \left[\frac{\theta_{x-\Delta x} - \theta_x}{\Delta x} - \frac{\theta_x - \theta_{x+\Delta x}}{\Delta x} \right] \tag{9.18}$$

If we look at Figure 9.2, it would appear that the solution variable for this equation should be θ_x, and this would be a reasonable assumption, for we are solving for the change in midpoint temperature for the central element, in terms of the temperature gradients across the element boundaries. This is, once again, really a central difference solution, for in reality:

- we will have, as a boundary condition, the value of temperature at the input end of the first element, and
- we would normally wish to solve for the temperature on the output end of the last element in the series.

It is sometimes easier to obtain a backward difference solution by solving for the temperature at the output end of each finite element. We shall see the application of this in an example below.

The above discretization with respect to the spatial variable, or discrete-space–continuous-time (DSCT) representation, is well suited to solution by means of general-purpose simulation facilities, from analog computers to simulation languages. The hardware constraints of analog computer implementation limit the sizes of finite-element models. Digital computer simulation imposes fewer such constraints, and DSCT techniques are widely used for distributed-parameter simulation.

The other discretization technique, that of continuous-space–discrete-time (CSDT) solution, is not much used. It is really more suited to hybrid computation where function storage and playback techniques make the method more feasible. We may, however, examine this method of working for the purposes of comparison with the DSCT solution.

Let us look at the CSDT discretization of equation (9.6) with reference to the difference equation (9.5). We obtain the ODE

$$\frac{d(\rho V)}{dx} = \frac{\rho_t - \rho_{t+\Delta t}}{\Delta t} \tag{9.19}$$

The theory behind this solution technique is that we start with the initial value for $\rho(x,t)$ as the spatial profile $\rho(x,0)$, assume a time increment Δt and generate a new spatial profile $\rho(x, \Delta t)$ by solving equation (9.19) using normal integration techniques. The new spatial profile is stored (in piecewise linear segments) as it is generated, and then used as input to equation (9.19) for solution at the next time increment. Repetitive solution of this equation on the analog part of a hybrid computer means that real-time solution can often be obtained. The nature of the solution also indicates that backward differencing with respect to time is the method to use for safety and stability.

The heat diffusion equation (9.14) may also be finite-differenced with respect to time (see equation (9.13)) to give

$$\frac{d^2\theta}{dx^2} = \frac{\rho C}{k}\left[\frac{\theta_{t+\Delta t} - \theta_t}{\Delta t}\right] \tag{9.20}$$

If we use the function storage and playback technique on a hybrid computer to solve this equation, then, once again, the nature of the solution indicates that backward differencing is best and safest.

9.3 STABILITY PROBLEMS

We have touched briefly on the stability problems associated with the solution of PDEs by finite-differencing techniques. The stability in question here has been that of the resulting ODEs themselves, and whether or not they would yield stable solutions. We now need to consider the question of numerical stability, given that the equations are themselves stable.

If we are carrying out a DSCT solution of a PDE, then the discretization will yield a set of ODEs with possibly a range of time constants (certainly true if the original PDE is non-linear). We may then invoke the conditions for numerical stability set out in Section 6.3 to ensure that the solution is numerically stable. In the same way as for normal ODEs, a well-behaved numerical integration algorithm operating under error control will be error limited, and as a consequence step sizes will be well within the limits for numerical stability. We shall examine, in a simulation example, the way in which the choice of discretization parameters can affect the time constants of the resulting ODEs, and the resulting effect upon numerical integration.

9.4 INITIAL AND BOUNDARY VALUES

In the discussion above, we have only briefly mentioned initial and boundary values or conditions. This facet of the definition of a distributed-parameter model is one that causes many problems for anyone trying to achieve a mathematically correct realization in terms of the portrayal of the model as a discretized system of ODEs.

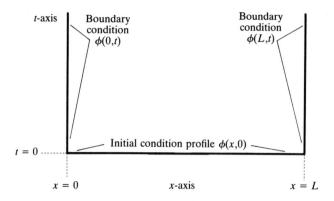

Figure 9.3   Frame of reference for solution of PDE in one spatial dimension.

Let us consider a distributed-parameter system of one spatial dimension such that $0 \leqslant x \leqslant L$. Figure 9.3 shows a two-dimensional schematic of the independent variables, such that time t is the ordinate and distance x is the abscissa. This is the frame of reference within which a PDE of one spatial dimension will operate, and we need to consider the initial and boundary values (or conditions) in terms of their location and extent within this framework. ODE solution with respect to time as the independent variable is fully characterized by the initial (point) values of the state variables. A partial derivative with respect to time in a PDE requires an equivalent specification to be made for a unique solution to be obtained. Since the state variables are now functions of time and distance the specification is somewhat different, for example

$$\text{ODE:} \frac{dw}{dt} = f(w,t) \tag{9.21}$$

where

$$w = w(t)$$

and the initial *value* of the state variable is given by

$$w|_{t=0} = w(0) \tag{9.22}$$

$$\text{PDE:} \frac{\partial \phi}{\partial t} + \frac{\partial \phi}{\partial x} = g(\phi, x, t) \tag{9.23}$$

where

$$\phi = \phi(x,t)$$

and the initial *profile* of the state variable is given by

$$\phi|_{t=0} = \phi(x,0) \tag{9.24}$$

which is the spatial profile that is defined between $x = 0$ and $x = L$ at time $t = 0$ (see Figure 9.3). However, since a spatial derivative is present in equation (9.23), an equivalent initial (or final) value needs to be defined at $x = 0$ (or $x = L$) in order to characterize the solution in the x-direction. This also takes the form of a 'profile' in time (i.e. a time transient) that describes the behaviour of the state variable as a function of time, either at the beginning or at the end of the spatial range of the solution. It is analogous to defining an initial or boundary value for a time-dependent state variable, and is also referred to as a *boundary condition*. It is defined as

$$\phi|_{x=0} = \phi(0,t) \tag{9.25}$$

or

$$\phi|_{x=L} = \phi(L,t) \tag{9.26}$$

as appropriate in terms of the requirements of the solution.

If we are going to solve equation (9.14), namely

$$\frac{\partial \theta}{\partial t} = \frac{k}{\rho C} \frac{\partial^2 \theta}{\partial x^2} \tag{9.14}$$

then because of the presence of the second derivative in x, we need an extra boundary condition to be defined. This can take the form of either boundary values at *both* ends of the spatial range (at $x = 0$ *and* at $x = L$), or a boundary value and a derivative. For example, for equation (9.14) we could have boundary conditions as either

$$\theta(0,t) \quad and \quad \theta(L,t)$$

(boundary values at both ends of the space), or

$$\theta(0,t) \quad and \quad \frac{d\theta(0,t)}{dx}$$

(initial values of state variable and derivative with respect to x), or any other suitable combination of initial and final values and derivatives.

If we now look at Figure 9.4, we have a schematic diagram of discretization with respect to distance. In this case, the spatial profile that constitutes the initial value of the function has been split up into a number of point initial values, one for each of the ODEs that have replaced the PDE in the discretization process. It now remains to consider the influence of the boundary conditions. If the boundary conditions are wholly initial valued (i.e. functions defined at $x = 0$), then it may be expedient to solve for values of the state variables at the ends of the finite elements, moving from left to right. If final-value boundary conditions are set, then solution from right to left is advocated. If a mixture of initial- and final-value boundary conditions are set, then the equation set becomes more difficult to solve, and will probably require iteration. In flow problems, it is usual to have a flow input end, at which boundary conditions may be specified. In such

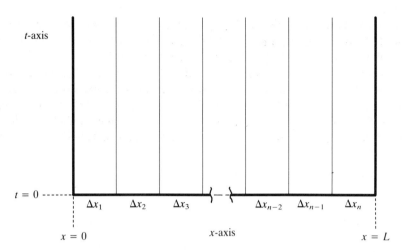

Figure 9.4 Frame of reference for one-dimensional discretization with respect to distance.

cases, difficulties with iteration are avoided. It is worth noting that in a case where initial and final conditions have been specified together, DAE solution can help in overcoming the 'closed' nature of the problem when it comes to achieving a solution. The iteration mentioned above is already encompassed in the Gear-type predictor–corrector integration schemes used for DAE solution. The possibility should not be discounted, however, that an 'index problem' could arise from this situation.

When we discretize with respect to time, the position is somewhat different. If we examine Figure 9.5, we can see that the initial spatial profile has to be set as a continuous function between $x = 0$ and $x = L$. The boundary conditions become a series of point values set at either $x = 0$ or $x = L$, as the occasion demands. These are then initial or final values for integration with respect to x at each time step.

Specification of boundary conditions is not usually too much of a problem, for in the majority of cases these will be defined as input or output functions of time. Initial-value spatial profiles, on the other hand, are much more difficult to specify and realize. It is often expedient to assume a basic form for a spatial profile at time $t = 0$, and allow the PDE solution to 'float' until the correct spatial profile is formed, consistent with the initial conditions required for the model at hand. This technique will be illustrated in a case study in Chapter 11.

9.5 SIMPLE SIMULATION EXAMPLE

Most of our simulation work so far has been with a view to producing time transients that illustrate the behaviour of the relevant variables within their

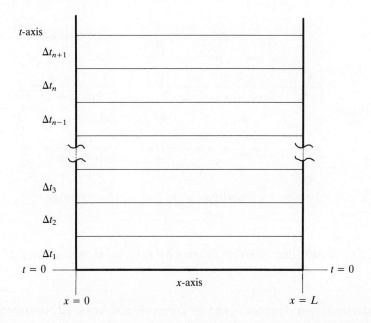

Figure 9.5 Frame of reference in one spatial dimension, for discretization with respect to time.

respective simulation models. The lumped-parameter modelling that has been implemented requires no information to be gathered about distributed effects such as the relationships of variables with respect to distance as well as time. In distributed-parameter system models, however, it often happens that the space- and time-independent variables are equally important in terms of the information content required from the simulation results.

It is universally accepted that a rapidly changing time-dependent variable in a digital simulation requires a smaller communication interval for output than one that varies only slowly, if the information content of the resulting time transient is to be preserved. This is probably equally true for a spatial profile that is output at a particular instant in time. The output intervals in terms of the spatial variable would need to be smaller where the profile shape can change rapidly, in order that such changes may be faithfully portrayed. Besides this, the communication interval for spatial profile output would need to be shorter in order that rapidly changing spatial profiles could be output more frequently.

In the time domain, due attention has to be paid to the stability of numerical integration methods which are finite-difference techniques with respect to time. This is done by using variable step integration algorithms which ensure numerical stability as well as controlling truncation error. The modelling of spatial characteristics by finite-difference approximations requires the corresponding care to be taken in the choice of finite-element sizes to ensure spatial stability.

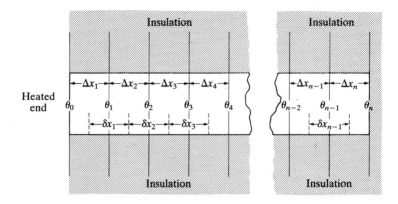

Figure 9.6   Spatial discretization scheme for solution of heated bar problem.

In order to try to illustrate some of the problems involved, consideration is given to a distributed-parameter system which is a function of time and one distance variable only, in keeping with the rest of this chapter. The method of solution adopted is the DSCT (Discrete-Space–Continuous-Time) approach whereby each PDE in the model is replaced by a set of ODEs, owing to discretization of the space variable. The resulting model is then manipulated to increase the information content from the simulation results.

The simulation example used for illustration purposes is the well-known 'heated bar' model in which a heated metal bar is fully covered by perfect insulation except at one end (see Figure 9.6). At time zero, the bar temperature is uniformly zero on a normalized scale. A step increase in temperature is applied at the uninsulated end, and the new level is maintained throughout the duration of the simulation. The PDE describing the ideal behaviour of the heat diffusion process is given by

$$\rho C \, \frac{\partial \theta}{\partial t} = k \, \frac{\partial^2 \theta}{\partial x^2} \tag{9.14}$$

based on a similar idea introduced by Franks (1967) and already introduced in Section 9.2, where

  ρ is the density of the material in the bar
  C is the specific heat of the material in the bar
  k is the thermal conductivity of the material in the bar
  θ is the temperature (in normalized units)
  x is the distance from the heated end of the bar
  t is time

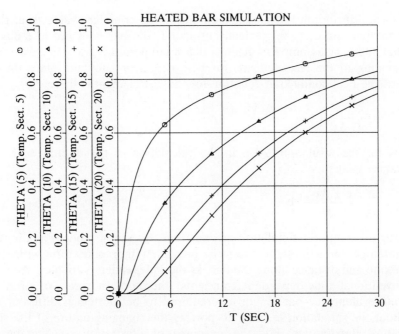

Figure 9.7   Heated bar simulation: time transients for temperatures at ends of four out of the 20 equally sized sections.

The initial-value spatial profile for this model is given by

$$\theta(x,0) = 0.0 \tag{9.27}$$

while the initial-value input time transient is given by

$$\theta(0,t) = 0.0 \quad (t < 0)$$
$$= 1.0 \quad (t \geqslant 0) \tag{9.28}$$

For the purposes of illustration, it is desirable to retain the ability to integrate with respect to time as the independent variable. The DSCT approach gives rise to a simple discretization scheme implemented over the length of the bar, but, as promised, it is different from that shown in Figure 9.2 and implemented in equation (9.18). Instead of being midpoint values, the discretized temperatures are set on the element boundaries as shown in Figure 9.6. The resulting set of equations is

$$\frac{\mathrm{d}\theta_i}{\mathrm{d}t} = \frac{k}{\rho C} \frac{1}{\delta x_i} \left(\frac{\theta_{i-1} - \theta_i}{\Delta x_i} - \frac{\theta_i - \theta_{i+1}}{\Delta x_{i+1}} \right) \quad (i = 1,2,3,\ldots,n-1) \tag{9.29}$$

for all elements except the last. This allows for easy implementation as θ_0 of the initial-value boundary condition given by equation (9.28). In this implementation, we do have difficulty in realizing a solution for the boundary value θ_n at the far end of the bar, since, being perfectly insulated, the bar does not lose any heat from that end. The assumption made is that a temperature gradient exists only at the beginning of the last element, for heat conduction to take place into that element. The heat conduction equation for the last element is therefore

$$\frac{d\theta_n}{dt} = \frac{k}{\rho C} \frac{1}{\Delta x_n} \left(\frac{\theta_{n-1} - \theta_n}{\Delta x_n} \right) \tag{9.30}$$

to allow for the insulation. Also in the calculation of the discretized form of temperature gradients

$$\delta x_i = \left(\frac{\Delta x_i + \Delta x_{i+1}}{2} \right) \tag{9.31}$$

In the digital computer simulation arising from this set of model equations, the solution temperature transients are to be presented in graphics form with respect to both time and distance along the bar, as the independent variables. As well as the conventional time-dependent transients, therefore, a series of profiles of temperature along the bar length are required to be plotted at different times throughout the simulation in order to portray the changing nature of the spatial temperature distribution. For the purposes of illustration, the temperature characteristics are first simulated for a heated bar 10 units in length, divided into 20 equal sections.

The simplest case to be considered is that where the spatial graphics are output at regular intervals of time. The duration of the simulation for all tests is fixed at 30 seconds, and for this first case, spatial graphics are output at time zero and every 3 seconds thereafter. Figure 9.7 shows the time transients of temperature obtained for four of the 20 sections. Figure 9.8 shows a complete set of time transients for all the sections together, superimposed on the same set of axes. In Figure 9.8 we can see how the elements near the heated end show the most rapid response to the step input of temperature θ_0, while those further away respond more slowly.

What we do want to see for this exercise is a plot of the spatial profiles as sampled every 3 seconds during the simulation run. Figure 9.9 shows these profiles plotted as temperature against distance from the heated end of the bar, for all 20 equally sized finite elements. This gives us much of the information about profile shapes that we wanted, but two things are amiss with these profiles:

- There is some 'aliasing' of the profile produced at time $t = 0$ which, in reality, should show an instantaneous reduction in temperature at time $t = 0$, from 1.0 to 0.0. This reduction is phased over the whole element length.
- A large amount of profile information is missing between time $t = 0$ and $t = 3.0$ (the first profile sample after the beginning of the run). The most

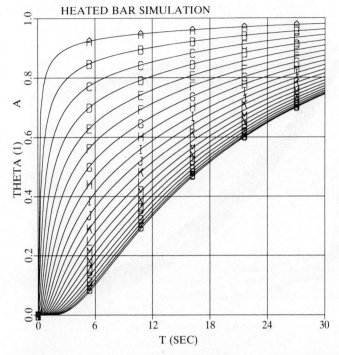

Figure 9.8   Heated bar simulation: time transients for temperatures at ends of all 20 equally sized sections.

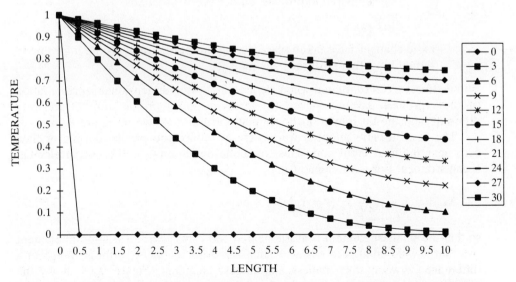

Figure 9.9   Heated bar simulation: spatial temperature profiles across equally sized sections, output at regular intervals of 3 seconds.

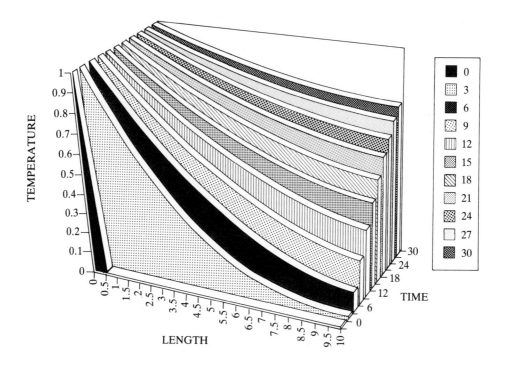

Figure 9.10 Heated bar simulation: three-dimensional representation of spatial temperature profiles, output every 3 seconds.

rapid changes have taken place during this period, and profile information should be gathered during this time.

These effects are accentuated when we look at a three-dimensional representation of these profiles. Evenly spaced (in terms of time) profile outputs are very different near the beginning of the simulation run, as shown in Figure 9.10.

The first measure adopted was to try to reduce the profile aliasing at time zero near the heat input end. This was done by adopting a differential element sizing strategy. For each element i,

$$\Delta x_i = i.L \bigg/ \sum_{i=1}^{n} i \quad (i = 1,2,3,\ldots,n) \tag{9.32}$$

so that the smallest element is at the heated end. This gives a variation in element size that is wide enough to ensure a small-sized first element, without too great a difference between the smallest and largest elements. Figure 9.11 shows the resulting temperature transients for four of the elements, while Figure 9.12 shows the complete set of temperature–time transients. Note the aliasing that occurs

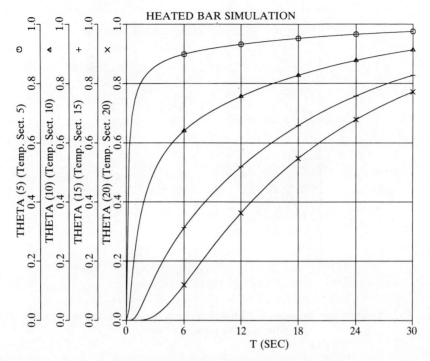

Figure 9.11   Heated bar simulation: time transients for temperatures at ends of four out of the 20 differentially sized sections.

now on the time transients for those elements near the heated end. The communication interval for time transients is not small enough for the rapid changes to be captured near time zero. Figure 9.13 shows a plot of the spatial profiles produced from this simulation run. The temperature aliasing at time zero has almost been eliminated by the small initial element size, but the missing profile information near time zero has not yet been remedied.

The next measure adopted to try to increase the information content of the simulation graphics output was to implement differential sample rates for the profile sampling action. It was considered that the sample rate should somehow be a function of time derivatives of the temperature, so more information on rapid changes would be captured early on in the simulation run. The compromise adopted was to have sample rate proportional to the average of all the time derivatives of temperature, by calculating an instantaneous sample period as

$$\Delta t_s = \frac{W}{\left(\sum_{i=1}^{n} \dot{\theta}_i \right) \bigg/ n} \tag{9.33}$$

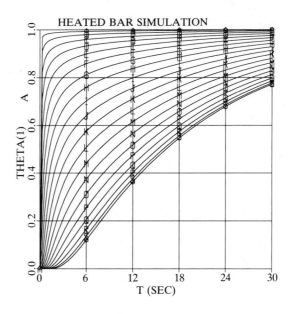

Figure 9.12   Heated bar simulation: time transients for temperatures at ends of all 20 differentially sized sections.

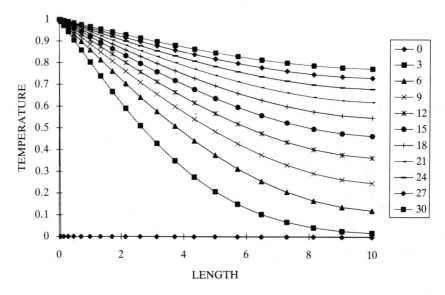

Figure 9.13   Heated bar simulation: spatial temperature profiles across differentially sized sections, output at regular intervals of 3 seconds.

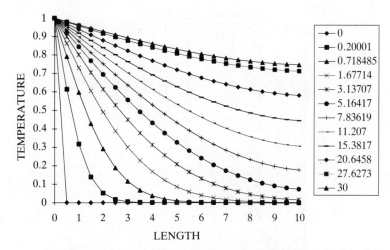

Figure 9.14 Heated bar simulation: spatial temperature profiles across equally sized sections, output at unequal intervals of time during the run.

where

$\dot{\theta}_i$ is the value of the time derivative of temperature for the ith element in the bar

W is a weighting factor applied to govern the total number of samples in a simulation run

Δt_s is the instantaneous sample period

This sampling was implemented by means of a *discrete* section in an ACSL run, whereby the sample period was calculated as a time to next event. In this way, the time interval between one profile output event and the next was inversely proportional to the mean value of all the time derivatives at the first event time. Time zero was set as the first event in the process. The time transients produced are very similar to those in Figures 9.7 and 9.8 if the finite elements are of equal size, but as you can see in Figure 9.14, there was much more spatial profile information produced in the first 3 seconds of the simulation run. The legend in Figure 9.14 gives the actual sample times during the run, from which you can see that three profile samples were obtained within this space of time.

The final test with this simulation was to carry out a run with differential element sizing together with derivative-based sampling action for the spatial profiles. Figure 9.15 shows the complete set of time transients for all the finite-element temperatures. Notice how the aliasing has disappeared from the transients just after time zero. The event-based profile sampling has helped to fill in time-dependent information at the beginning of the run as well as providing better spatial profile information. Figure 9.16 shows the spatial temperature profiles from this last run. The legend shows just how frequent the sampling was at the beginning of the run, accentuated by the high temperature-derivative values

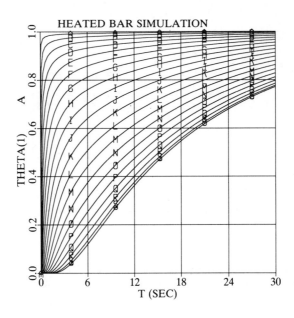

Figure 9.15 Heated bar simulation: time transients for temperatures at ends of all 20 differentially sized sections, with spatial temperature profile sampling at unequal intervals.

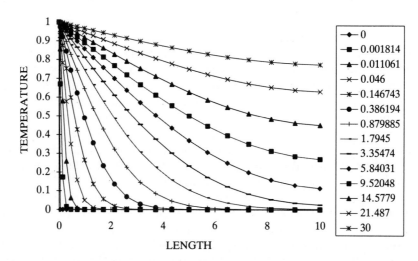

Figure 9.16 Heated bar simulation: spatial temperature profiles across differentially sized sections, output at unequal intervals of time during the run.

Table 9.1 Typical run-timing data for heated bar problem

Time and space discretization modes		Execution times (seconds)	
Finite-element spacing	Spatial profile output spacing	Runge–Kutta–Fehlberg (fifth order)	Gear's method (variable order)
Regular	Regular	8.6	13.8
Differential	Regular	207	19.0
Regular	Event based	8.2	14.0
Differential	Event based	207	17.5

obtained in the smallest elements. Note, too, that the early spatial profiles, although showing spatially very steep rates of change, are not much aliased by the finite differencing, since the small finite-element sizes near the heated end of the bar give a better portrayal of rapid changes in much the same way as small time steps do for time transients.

This brings us on to a feature of differential element sizing of which you should be aware. If you look at equation (9.29), by inspection you should notice that the time constant of the equation will be given approximately by

$$\frac{1}{\tau} \approx \frac{k}{\rho C} \frac{1}{\Delta x} \frac{1}{\delta x} \tag{9.34}$$

and will therefore be affected by the values of Δx and δx. The smaller these values, the smaller will be the time constant of the equation concerned, but since the spatial effects are second order, the time constant varies roughly as the square of the element size. In the adoption of differential element sizing, therefore, a wide-band eigenvalue problem is created, and the equations become stiff. Table 9.1 shows typical run-timing data obtained from the use of a fifth-order Runge–Kutta–Fehlberg (RKF) algorithm and Gear's variable-order method for stiff systems, when applied to the heated bar problem. The RKF integration algorithm coped better than Gear's method in the execution of those simulation runs where the finite elements were all of equal size. However, when the finite elements of the bar were differentially sized, the stiffness of the model equations caused long solution times with RKF. Gear's method took only marginally longer.

These methods for portraying both time transients and spatial profiles can be extended readily to other distributed-parameter system models. We shall see in Chapter 11 how the methods are applied to a more complicated problem.

9.6 REFERENCES

BEKEY, G A and KARPLUS, W A (1968) *Hybrid computation*, New York: John Wiley, pp. 211–43.

FRANKS, R G E (1967) *Mathematical modeling in chemical engineering*, New York: John Wiley, pp. 228–32.

HIMMELBLAU, D M and BISCHOFF, K B (1968) *Process analysis and simulation: deterministic systems*, New York: John Wiley.

HUGHES, W F and GAYLORD, E W (1964) *Basic equations of engineering science*, Schaum's Outline Series, New York: McGraw-Hill.

SCHUCHMANN, H (1970) 'On the simulation of distributed-parameter systems', *Simulation*, vol. 14, pp. 271–9.

10

Case Study: Simulation of Direct Digital Computer Control

10.1 INTRODUCTION

Chapter 8 gave us an introduction to techniques that we may use for the digital computer simulation of sampled-data systems. It soon became clear for this area of simulation modelling that event processing as used in discrete-event simulation was of great value for handling the discontinuities that feature prominently. This particular chapter concerns a case study that not only is an illustration of sampled-data control with the periodic updating of control parameters, but includes a great deal of discrete-event activity in the control actuation.

In the United Kingdom, reactor technology for the nuclear power industry has, with few exceptions, followed the gas-cooled reactor path. The earlier nuclear power station reactors were of the 'Magnox' type, which were followed by the so-called 'Advanced Gas-cooled Reactors' (AGRs) designed to run at a higher temperature than their predecessors. The case study presented in this chapter is the simulation of a much simplified model of part of an AGR control system.

In a nuclear reactor, the nuclear fission occurring in the fuel produces large numbers of neutrons that move around at high speed in a so-called *neutron flux*. Some get 'captured' by fissionable atoms (e.g. ^{235}U) in the fuel, which then disintegrate into fission products and more neutrons. If the neutron flux is just high enough, the fission process sustains itself in a series of chain reactions, and the reactor becomes *critical*. Any higher neutron flux results in the reactivity growing until some other factors again bring it into balance. These factors are the reactivity–temperature characteristics of the fuel, moderator, and other reactor components, of which the control rods are the most important example, since they damp down the reaction by absorbing excess neutrons. The role of a *moderator* in a nuclear reactor is to slow down the fast neutrons by a series of collision processes, absorbing their kinetic energy and, in effect, converting it into heat energy.

Gas-cooled nuclear reactors of this type employ a large mass of graphite as the moderator. They are designed essentially to run at reasonably constant temperature, or at least with very slow changes of temperature. Accordingly, they

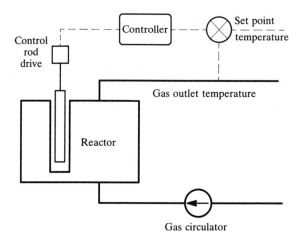

Figure 10.1   Schematic diagram of AGR gas outlet temperature control system.

are best suited to base-load situations, where there is little or no requirement to respond to changes in electricity demand over a supply grid.

Figure 10.1 shows a schematic diagram of this simplified temperature control system. The control loop is one of many within the reactor whereby the reactor gas outlet temperature is regulated. A digital control computer, by scanning all the loops in succession, computes for each loop any necessary control action to be actuated by a control rod drive. The control philosophy for each loop is quite simple. Any error between the temperature set point and the measured gas outlet temperature for a loop causes the controller (i.e. the digital control computer) to determine a course of control action. This results in raising or lowering the corresponding control rod, thereby increasing or decreasing control reactivity within that portion of the core around the control rod, to try to reduce the error in local gas outlet temperature.

Control actuation therefore consists of control rod displacement within the reactor core, in response to control system demands. A large temperature error will demand a correspondingly large control rod displacement to restore temperature control. For normal actuation, this would result in a faster actuator motor speed being set in by the controller. For the system used in this model, however, the control rod drives are constant speed devices, and therefore variation in control demand has to be satisfied in a different way.

Since a large number of control rod drives all operate under the supervision of a single control computer, each of the drives is 'visited' regularly by this computer to sample its gas outlet temperature and to compute controller demands. This constitutes sampled-data control as anyone would understand it, the sampling period of which is related to the total time to scan all the loops. For a constant speed control actuator (i.e. the control rod drive motor) control

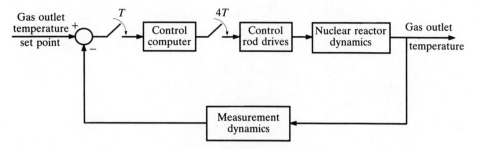

Figure 10.2   Block diagram of multirate sampled-data SISO system used for AGR gas outlet temperature control.

variation comes about by running the drive over a varying fraction of the sampling period to give so-called *mark/space actuation*. Some further explanation below should make this concept clearer.

Figure 10.2 shows a block diagram for the simple single-input–single-output (SISO) sampled-data control system used for this example. This is a multirate sampled-data system in that the temperature error and control computer output are sampled at different rates, usually one an integer fraction or multiple of the other. In this example the choice was to sample the temperature error four times faster than the control output. While it appeared to be sufficient to update the control actuation every 8 seconds, for various reasons it was better to sample the temperature error every 2 seconds. As we shall see later, the control actuation demanded is computed in terms not only of temperature error as just sampled. It is also a function of error samples taken during previous 'visits' from the control computer. If these 'visits' are closely spaced, that is the sampling period for temperature measurement is small, past information stored is less out of date. This would not be the case for temperature sampling at the slower rate.

10.2 A LUMPED-PARAMETER MODEL OF THE GAS TEMPERATURE CONTROL SYSTEM

Figure 10.3 shows a cut-away schematic diagram of a single gas flow channel in an AGR. Within this gas space is fixed a fuel element consisting of a fuel pin contained in an alloy cladding or 'can'. Cool gas at temperature Tg_{in} enters the gas space at the bottom of the reactor core, and flows up the annular space between the fuel element and the moderator, abstracting heat from both of these on its passage upwards. The temperature of the gas emerging from the top of the core, Tg_{out}, is used to control the reactivity within the core. This affects the amount of heat generated in the fuel and moderator, and transferred to the gas. The actual situation within an AGR is much more complicated than this. Multiple

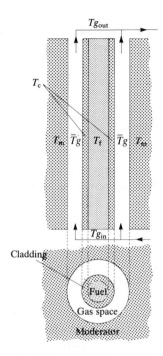

Figure 10.3   Cut-away schematic of a single AGR gas flow channel.

gas flow paths exist to obtain maximum heat transfer from the core to the cooling gas, but the simple model used in this case study and outlined below serves to illustrate the principles involved. The distributed nature of the reactor dynamics has also been ignored and this localized control system treated as a lumped-parameter system operating in isolation from all others of its type.

10.2.1 Model Equations

With reference to the diagram in Figure 10.3, the following equations are given for the simple lumped-parameter AGR temperature control model that is the subject of this case study:

Change in fuel reactivity

$$\Delta\rho_F = 816.8(T_{F_0} - T_F) \tag{10.1}$$

where

$\Delta\rho_F$ denotes change in reactivity due to fuel temperature
T_{F_0} is the reference fuel temperature (initial value in this case)
T_F is current fuel temperature

Change in moderator reactivity

$$\Delta\rho_M = 770.0(T_M - T_{M_0})$$ (10.2)

where

$\Delta\rho_M$ denotes change in reactivity due to moderator temperature
T_M is current moderator temperature
T_{M_0} is the reference moderator temperature (initial value)

Reactor power contribution from fission product groups

$$\frac{dP_{2_j}}{dt} = \varepsilon_j(a_j P_1 - P_{2_j})$$ (10.3)

where

P_{2_j} is the power contribution from the jth fission product group
ε_j is the inverse time constant for the release of fission product power from the jth fission product group
$\boldsymbol{\varepsilon} = [8 \times 10^{-6} \ 5.5 \times 10^{-4} \ 6.96 \times 10^{-3} \ 0.0417 \ 0.221 \ 0.475] a_j$
   is the fraction of total fission product power contributed by the jth group
$\mathbf{a} = [0.175 \ 0.218 \ 0.184 \ 0.205 \ 0.111 \ 0.107]$
P_1 is the reactor power from fission

Total power from fission product groups

$$P_2 = \sum_{j=1}^{6} P_{2_j} \quad (j = 1, \ldots, 6)$$ (10.4)

Fraction of heat generated in fuel

$$H_F = 0.8701 P_1 + 0.0649 P_2$$ (10.5)

Fraction of heat generated in moderator

$$H_M = 0.0434 P_1 + 0.0216 P_2$$ (10.6)

Fuel temperature

$$\frac{dT_F}{dt} = 0.0571 H_F - 0.2399(T_F - T_C)$$ (10.7)

where

T_C is the temperature of the 'can' (fuel cladding)

Total reactivity change within reactor

$$\Delta\rho = \Delta\rho_F + \Delta\rho_M + \Delta\rho_D + \Delta\rho_C$$ (10.8)

where

$\Delta\rho_D$ is a reactivity disturbance introduced into the system
$\Delta\rho_C$ is a reactivity change introduced by the control system

Reactor power from delayed neutrons

$$\frac{dD_i}{dt} = \beta_i P_1 - \lambda_i D_i \quad (i = 1, \ldots, 6) \tag{10.9}$$

where

β_i are delayed neutron fractions for six groups
$\boldsymbol{\beta} = [0.00021 \ 0.00117 \ 0.00104 \ 0.00224 \ 0.00070 \ 0.00014]$
λ_i are inverse time constants for delayed neutrons
$\boldsymbol{\lambda} = [0.0127 \ 0.0317 \ 0.115 \ 0.311 \ 1.4 \ 3.87]$

Reactor power from fission

$$0.001\frac{dP_1}{dt} = P_1 \left(\frac{\rho}{10^5} - \bar{\beta} \right) + \sum_{i=1}^{6} \lambda_i D_i \tag{10.10}$$

where

ρ is the total reactivity in the system
$\bar{\beta}$ is the sum of all the delayed neutron fractions (0.0055)

Average gas temperature across reactor

$$\bar{T}_G = \frac{0.7161 M_G^{0.8} T_C + 0.1658 M_G^{0.8} T_M + 0.8572 M_G T_{G_{in}}}{0.8572 M_G + 0.8819 M_G^{0.8}} \tag{10.11}$$

where

M_G is the gas flow through the fuel element channel
$T_{G_{in}}$ is the inlet gas temperature at the bottom of the channel

Moderator temperature

$$\frac{dT_M}{dt} = 0.002344 H_M - 0.002619 M_G^{0.8}(T_M - \bar{T}_G) \tag{10.12}$$

Gas temperature at reactor channel outlet

$$T_{G_{out}} = 2.0\bar{T}_G - T_{G_{in}} \tag{10.13}$$

Measured gas outlet temperature

$$2.5 M_G^{-0.75} \frac{dT_{G_M}}{dt} = T_{G_{out}} - T_{G_{in}} \tag{10.14}$$

Fuel 'can' temperature

$$\frac{dT_C}{dt} = 1.8078(T_F - T_C) - 1.7466M_G^{0.8}(T_C - \bar{T}_G) \qquad (10.15)$$

To assess the capability of the control system for regulation, we have to set up equations for the disturbances with which to test the model. In a test of the gas outlet temperature control, two types of disturbance are likely to provide a severe test of the control system, namely a step disturbance in reactivity, and a step disturbance in coolant gas flow. Each of these is likely to cause a severe transient in gas outlet temperature, with which the control system would have to cope.

Step change in reactivity

$$\Delta\rho_D = 0 \qquad t < t_{RD}$$
$$\Delta\rho_D = -20 \qquad t \geqslant t_{RD} \qquad (10.16)$$

where

t_{RD} is the time of application of the reactivity disturbance.

Step change in gas flow

$$M_G = 1.0 \qquad t < t_{GD}$$
$$M_G = 0.9 \qquad t \geqslant t_{GD} \qquad (10.17)$$

where

t_{GD} is the time of application of the gas flow disturbance.

At this point, we need to consider the method of application of control reactivity to the system. We have already come across the term mark/space. The time has come to examine this mode of control actuation and how we apply it to the regulation of the AGR gas outlet temperature.

For the purposes of this case study, a simple lead–lag compensator is implemented digitally to provide the necessary control action. We may describe this best in Laplace transform notation.

Compensator transfer function

$$\frac{U(s)}{E(s)} = G_C(s) = K \frac{(1 + T_1 s)(1 + T_2 s)}{(1 + T_3 s)(1 + T_4 s)} \qquad (10.18)$$

where the compensator time constants are given as

$$T_1 = 1.0, \quad T_2 = 1.0, \quad T_3 = 0.05, \quad T_4 = 20.0$$

$U(s)$ is the Laplace transform of the control output
$E(s)$ is the Laplace transform of the measured temperature error
K is the control gain

If we define a sampling period T_s (in this case, the 'space') it may be shown that the following digital controller algorithm is equivalent to the continuous compensator given in equation (10.18) above.

Digital controller algorithm

$$U_N = K \frac{A_1}{A_2} \left(E_N - \frac{B_1}{A_1} E_{N-1} + \frac{C_1}{A_1} E_{N-2} \right) + \frac{B_2}{A_2} U_{N-1} - \frac{C_2}{A_2} U_{N-2} \qquad (10.19)$$

where

$$A_1 = \frac{(T_s + T_1).(T_s + T_2)}{T_s^2} \qquad (10.20)$$

$$A_2 = \frac{(T_s + T_3).(T_s + T_4)}{T_s^2} \qquad (10.21)$$

$$B_1 = \frac{(T_1 + T_2) T_s + 2 T_1 T_2}{T_s^2} \qquad (10.22)$$

$$B_2 = \frac{(T_3 + T_4) T_s + 2 T_3 T_4}{T_s^2} \qquad (10.23)$$

$$C_1 = \frac{T_1 T_2}{T_s^2} \qquad (10.24)$$

$$C_2 = \frac{T_3 T_4}{T_s^2} \qquad (10.25)$$

and

U_N is controller output at present time
U_{N-1} is controller output at last sampling time
U_{N-2} is controller output at penultimate sampling time
E_N is temperature error at present time
E_{N-1} is temperature error at last sampling time
E_{N-2} is temperature error at penultimate sampling time

If the control rod drive speed were continuously variable, the demanded speed would be given by

$$\dot{\rho}_C = U_N \qquad (10.26)$$

However, since the drive speed is fixed at the maximum rate $\dot{\rho}_{max}$, we may obtain variations only by applying the *mark/space ratio* to give

$$\frac{M}{S} \dot{\rho}_{max} = U_N \qquad (10.27)$$

from which we obtain the following expression for the demanded 'mark':

$$M = \frac{S}{\dot{\rho}_{max}} U_N \qquad (10.28)$$

If, for example as in this model, we have a 'space' S of 8 seconds, a maximum rate of reactivity change $\dot{\rho}_{max}$ of 1 unit/s (normalized reactivity units) and a demanded rate of reactivity change U_N of 0.75 units/s, then the demanded 'mark' M will be 6 seconds.

Control reactivity is added to, or subtracted from, the total reactivity within the reactor core by driving control rods up or down, to withdraw them from, or insert them into, the core. The control rods are neutron absorbers; hence greater insertion into the core reduces the overall reactivity present in terms of neutron flux. For our purposes, the control rod movement into or out of the reactor core is represented by reactivity change.

The rate of movement of the control rod drives is constant, once they reach maximum velocity. Movement upwards takes the form of an initial acceleration to maximum velocity. This is followed by constant velocity for a specified period, and then by deceleration to standstill. Movement downwards may be described in the same way. The difference between the two directions of movement lies in the different acceleration and deceleration times that apply to each because of the effect of gravity upon rod movement.

Figure 10.4 illustrates the movement of a control rod in terms of rates of control reactivity change. For one control sampling period (the 'space'), the rod moves at maximum rate for some fraction of the 'space' that we term the 'mark' ($0 \leqslant$ mark/space $\leqslant 1.0$). The control computer determines the value (positive or negative) that the 'mark' should have for a particular control sampling period. The result is converted into control rod actuation as shown by the positive and negative reactivity rate timing diagrams in Figure 10.4. At the beginning of a 'space', once the value and sign of the corresponding 'mark' have been determined, the control rod accelerates from standstill to its maximum rate. The time to accelerate upwards or downwards is shown by the '4' at various points on the diagram. Constant rod speed is maintained until a time from the beginning of the sampling period corresponding to the 'mark' has elapsed (shown as '3' in the diagram). Deceleration takes place until the rod again reaches standstill. Several unusual situations may arise to modify the behaviour of the control rod. These are listed below:

- The rod may not reach standstill before the end of the 'space'. In this case, acceleration takes place away from the final velocity at the end of the previous 'space' (see first and second periods in the negative reactivity timing diagram).
- If the acceleration time is greater than the required 'mark', the rod will not reach maximum velocity (as shown by the second period in the positive reactivity timing diagram).

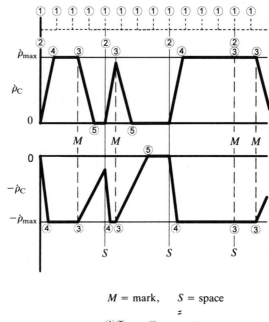

Figure 10.4   Mark/space control rod actuation shown as changes in control reactivity.

- If the demanded 'mark' exceeds the 'space', the rod will still be travelling at maximum velocity at the end of the 'space'. Control action in the next 'space' will start at maximum velocity without the initial acceleration (see the third period in the timing diagrams for both positive and negative reactivity).

We may express these concepts mathematically as follows:

(a) $(M > 0) \wedge (t < M)$

$$\frac{d\dot{\rho}_C}{dt} = +a_U \quad (\dot{\rho}_C < \dot{\rho}_{max}) \tag{10.29}$$

$$\frac{d\dot{\rho}_C}{dt} = 0 \quad (\dot{\rho}_C \geqslant \dot{\rho}_{max}) \tag{10.30}$$

where

$\dot{\rho}_C$ is the rate of change of control reactivity
a_U is the acceleration in a positive direction (derived from control rod acceleration upwards)

(b) $(M > 0) \wedge (t \geqslant M)$

$$\frac{\mathrm{d}\dot{\rho}_\mathrm{C}}{\mathrm{d}t} = -d_\mathrm{U} \;\; (\dot{\rho}_\mathrm{C} > 0) \tag{10.31}$$

$$\frac{\mathrm{d}\dot{\rho}_\mathrm{C}}{\mathrm{d}t} = 0 \quad\;\; (\dot{\rho}_\mathrm{C} \leqslant 0) \tag{10.32}$$

where

d_U is the deceleration in a positive direction (derived from control rod deceleration upwards)

(c) $(M < 0) \wedge (t < |M|)$

$$\frac{\mathrm{d}\dot{\rho}_\mathrm{C}}{\mathrm{d}t} = -a_\mathrm{D} \;\; (\dot{\rho}_\mathrm{C} > -\dot{\rho}_{max}) \tag{10.33}$$

$$\frac{\mathrm{d}\dot{\rho}_\mathrm{C}}{\mathrm{d}t} = 0 \quad\;\; (\dot{\rho}_\mathrm{C} \leqslant -\dot{\rho}_{max}) \tag{10.34}$$

where

a_D is the acceleration in a negative direction (derived from control rod acceleration downwards)

(d) $(M < 0) \wedge (t \geqslant |M|)$

$$\frac{\mathrm{d}\dot{\rho}_\mathrm{C}}{\mathrm{d}t} = +d_\mathrm{D} \;\; (\dot{\rho}_\mathrm{C} < 0) \tag{10.35}$$

$$\frac{\mathrm{d}\dot{\rho}_\mathrm{C}}{\mathrm{d}t} = 0 \quad\;\; (\dot{\rho}_\mathrm{C} \geqslant 0) \tag{10.36}$$

where

d_D is the deceleration in a negative direction (derived from control rod deceleration downwards)

At every control sampling instant (every 'space' in this context), the control computer updates the value of the demanded 'mark'. Therefore, irrespective of the state of control actuation at the end of the previous 'space', at the beginning of a new 'space' control actuation will proceed according to the above rules. Incorporation of control actuation in the model by means of the above equations should make its behaviour similar to that expected from the real process. In other words, we can feel sure that we are producing a reasonably realistic model of the control system.

10.3 RUNNING THE SIMULATION

The model as described above was programmed in the ACSL simulation language to run under Microsoft WINDOWS. The listing for the model description as written in the ACSL language is shown in the Appendix. This listing is fully commented in an effort to clarify the meaning of all the statements, and the values used for the model parameters are described at the top of the program.

The 'command module' that goes with the program to implement this model has been written to produce a series of time plots for different system variables. The first is a run-time plot for certain key variables, followed by post-run plots for several others. These are used as the basis of all the figures shown and discussed below.

Now that we have a model which we consider to be a fairly realistic portrayal of the real system, we need to subject it to a programme of testing to verify that it does indeed behave as we would expect.

For the purposes of this case study, it has been chosen to examine three facets of the model behaviour:

- The ability of the model to reach a steady state. This test requires a simulation run without any external stimuli to excite transient behaviour. The model is run from a given set of initial conditions and allowed to settle down to a steady state, if indeed it is capable of so doing. This model is expected to approach a steady state, even if it does not quite achieve this condition during the simulation timescale allowed.
- The time response of the model to a disturbance in coolant gas flow. The gas outlet temperature control system is expected to be able to cope with the temperature transient that follows this disturbance.
- The time response to a reactivity disturbance. A change in the power levels within the reactor causes alterations in the heat produced within the fuel and moderator. Again the control system under test should be able to cope with the resultant temperature transients.

When we carry out each of these tests, we are all the time looking for evidence that the model itself is behaving as we would expect, right down to the smallest detail. Whenever we carry out a run, we need to examine the resulting transient responses very carefully to be sure that everything is in order. This example should illustrate the principles involved.

10.3.1 Steady-state Properties of the Model

With the initial conditions as shown in the listing, we may carry out a simulation run to establish whether or not the model is capable of settling down to a steady state (i.e. whether or not it is *stable*). It often happens that the initial conditions chosen are not quite right for the achievement of a steady state. This does not

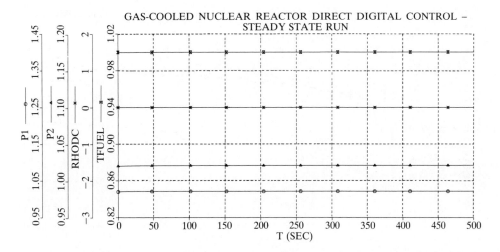

Figure 10.5   Steady-state test on AGR temperature control model: run-time plot.

matter, for what is really important is that a model does move towards a steady state, even if it is never quite reached.

Figure 10.5 shows a 500 second time transient plot obtained for four key variables, when our model is released from its initial conditions and allowed to 'float'. The variable 'RHODC' ($\dot{\rho}_C$) shows that no control action is taken during the run, while 'P1' and 'P2' (reactor power levels) and 'TFUEL' (fuel temperature for the loop) indicate that conditions within the reactor are more or less steady.

If we plot the same variables as in Figure 10.5 over their full ranges, we obtain the transients shown in Figure 10.6. Here we note that what looks like steady state on an enlarged scale is in reality not the case. However, we may exercise our judgement to decide that the system is sufficiently in equilibrium for our requirements, compared with the magnitude of the disturbances that we are going to apply.

Examination of the 'P1' transient in Figure 10.6 reveals a 'noisy' curve. Equation (10.10) contains a positive-feedback subsystem for reactor power, with the delayed neutrons alone providing some stabilization. The short time constant for this equation may also be expected to cause some problems for digital computer integration algorithms, so it is not surprising that some local instability is apparent in the solution. We can live with that, as long as the overall system behaviour is acceptable.

A further observation that we may make from examination of Figure 10.6 is that the fuel temperature (TFUEL) appears to be subject to some 'drift' upwards. This is more apparent towards the end of the transient when the system should be closer to the equilibrium condition. If we plot three other system temperatures and the actual gas temperature error, we obtain the results shown in Figure 10.7.

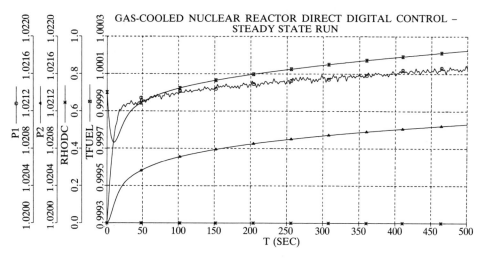

Figure 10.6 Steady-state test on AGR temperature control model: transients plotted over full ranges.

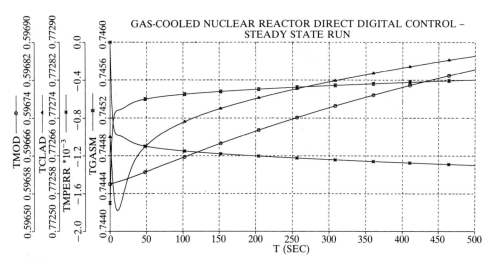

Figure 10.7 Steady-state test on AGR temperature control model: some temperature transients plotted over full ranges.

Here too, it may be interpreted that the system is subject to slight drift. The reactivity/temperature characteristics of the system are such that the true equilibrium point is not able to be reached within the timescale chosen for the simulation. The simulation results also show, however, that we are close enough to this point not to be too concerned. At this stage we are now in a position to proceed to our transient testing.

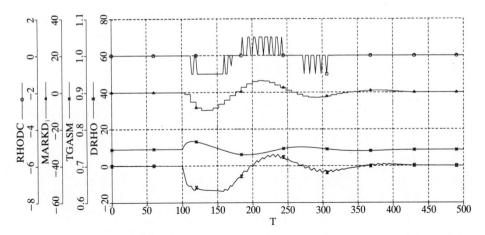

Figure 10.8 Model response to gas flow disturbance: event-driven control system; integrator-driven rod actuation; run-time plot.

10.3.2 System Response to a Disturbance in Coolant Gas Flow

The first test that we shall apply to the system is a gas flow disturbance. There is no particular reason for choosing this first, and we could just as easily have applied a reactivity disturbance instead. Figure 10.8 shows the run-time plot obtained for this test. One hundred seconds of simulation time are allowed to elapse before the disturbance is applied. The overall response to the disturbance is fairly good in that the system returns to something close to its original state after the transients have died away. This shows at least that the control system is doing its job.

When we get down to detailed examination of the transients, however, it soon becomes apparent that all is not well. For instance, if we look at the very first manifestation of control actuation after application of the disturbance, a small value of MARKD results in no control reactivity change whatsoever, over that particular 'SPACE'. Also, the acceleration and deceleration rates (both upwards and downwards) appear to be more or less the same, judging by the slopes on the 'crenellations' that characterize the plot for RHODC. This is at odds with what was programmed into the model.

Figure 10.9 shows an expanded plot for the same variables with the exception that measured gas outlet temperature ('TGASM') has been replaced by the continuous temperature error ('TMPERR'). Only a part of the overall plot is shown (between 100 and 300 seconds of simulation time). The magnified transients show further irregularities in the shapes of the 'peaks' for RHODC, indicating perhaps that some form of 'aliasing' (distortion) of the results has taken place. Apart from that, the control action looks plausible, in that the shape of

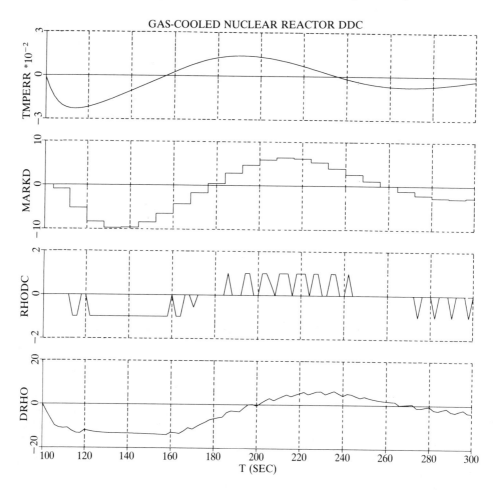

Figure 10.9   Model response to gas flow disturbance: event-driven control system; integrator-driven rod actuation; control action over timescale of 100–300 seconds.

MARKD follows the temperature error, and this results in an overall reactivity transient DRHO that also looks similar in shape.

When looking for clues that will lead us to the cause of the problem, it is often useful to amplify further a portion of the plot, so that we may examine it in detail. Figure 10.10 is a further amplification of that portion of the transients that lie between 100 and 200 seconds of simulation time. The plotting grid is divided into units of 10 seconds, and it is this that gives us a clue to the problem. If we examine the transient for DRHO, we can see that any changes of direction that occur always take place at intervals of 2 seconds. What is more, any acceleration and deceleration between full reactivity rate and zero is always accomplished in 2

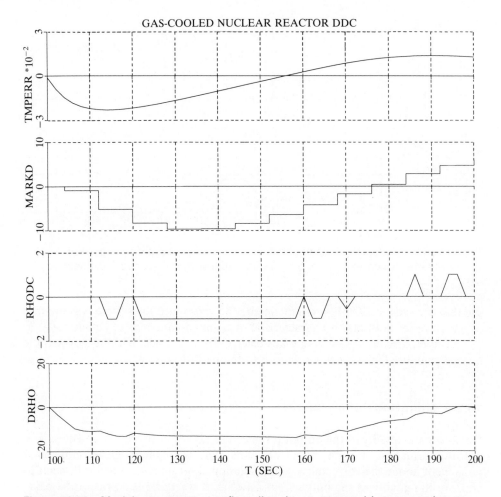

Figure 10.10 Model response to gas flow disturbance: event-driven control system; integrator-driven rod actuation; control action over timescale of 100–200 seconds.

seconds. From the figures used in the simulation for acceleration and deceleration rates, the times for full acceleration and deceleration should all be less than 1 second. This indicates that we are only seeing output that is plotted at 2-second intervals, and that much information is being lost in between.

Examination of the program listing at the point where *communication interval* is specified (CINTERVAL near the end) shows us that output was asked for every 20 seconds of simulation time, which in this example is far too infrequent. In fact, we obtain output every 2 seconds because it has been requested within the DISCRETE sections of the model that handle the multirate sampled-data control implementation. If you look at these sections within the program listing, you will

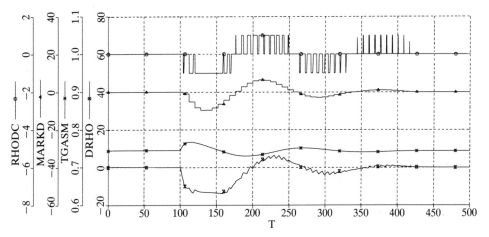

Figure 10.11   Model response to gas flow disturbance: event-driven control system; integrator-driven rod actuation; run-time plot with output every 0.1 second.

see that the calls to LOGD are implemented to reproduce the square-wave nature of a zero-order hold that is characteristic of sampled-data control action.

Temperature error sampling takes place every 2 seconds, and this governs the highest rate at which results are output to the plotting routines.

10.3.3 Improving Output Quality

Obviously the quality of the output produced so far is not satisfactory. There is so much information missing from the plots that we cannot make a detailed judgement about the correctness of the solution. Fortunately, it is not difficult to improve the quality of the output, but as we shall see there is a price to be paid.

It is a simple matter to reduce the communication interval to a value for which we could expect the output to look more authentic. If we choose a value of 0.1 second, we may implement this at run time by noting the name of the variable used to store the value (CINT in this case) and change it to 0.1 by typing as an interactive command at run time 'SET CINT=0.1'.

Running the problem again with all other parameters unchanged yields the run-time plot shown in Figure 10.11. If we compare this with Figure 10.8, we can see that the transients for MARKD, TGASM, and DRHO are substantially the same in both cases, indicating that output 'aliasing' is not really much of a problem for those variables. Where we do notice the difference, however, is in the respective shapes of the transients for RHODC. In Figure 10.11, where MARKD has a non-zero value, there is control action within the corresponding 'SPACE'. The resulting acceleration/deceleration peaks have far greater slopes that are more in line with what was programmed.

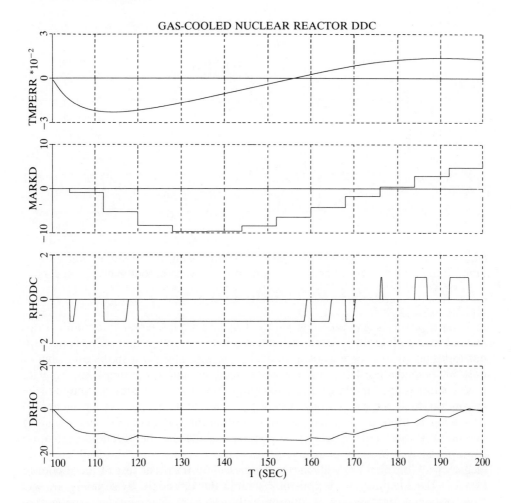

Figure 10.12 Model response to gas flow disturbance: event-driven control system; integrator-driven rod actuation; control action over timescale of 100–200 seconds with output every 0.1 second.

If we examine in Figure 10.12 a repeat of the plots of Figure 10.10, we can see the difference in the plots for RHODC. The distorted control action transients are replaced by peaks of control activity whose shape appears correct in every detail, down to the differences in acceleration/deceleration slopes for upward and downward movement of the control rods.

The price that we have to pay for this improvement in quality is the longer time taken to generate the solution. We are asking for a much smaller communication interval (0.1 second compared with the default value of 2 seconds produced by error sampling). The integration step size will therefore automatical-

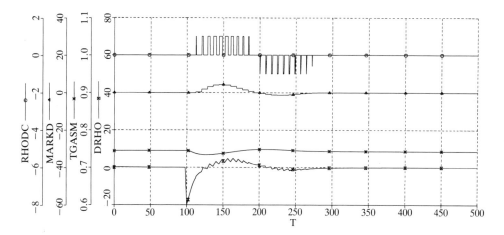

Figure 10.13   Model response to reactivity disturbance: event-driven control system; integrator-driven rod actuation; run-time plot with output every 0.1 second.

ly be kept at or below this new value. The larger number of integration steps needed to complete the solution will mean a longer solution time. This is perhaps not too much of a problem for a model of this size and for a simulation time of only 500 seconds. For larger models and longer simulation run times, solving with a very short output interval simply to obtain 'non-aliased' transients may not be an option that we can readily accept. We would like some other means of producing accurate transients with faster solution times to avoid the time penalties described above. Fortunately it is possible to extend the discrete simulation concepts used to implement sampled-data control to overcome this problem in an elegant way, but first we will carry out our second transient test of the model to establish the adequacy of the control system in this direction. By so doing, we will have obtained two sets of responses with which to compare any changes in modelling technique.

10.3.4 System Response to Reactivity Disturbance

Having evolved a strategy for obtaining accurate transient responses from our model, we are now in a position to test the model with the second of our proposed disturbances. If we apply a negative step change to the overall reactivity of the system, we would expect a drop in the gas outlet temperature. Figure 10.13 shows a run-time plot of the consequences of injecting a reactivity disturbance of −20 units into the model. The expected decrease occurs in the measured gas outlet temperature, and positive control action takes place to compensate with increased control reactivity. The system response to such a severe disturbance is

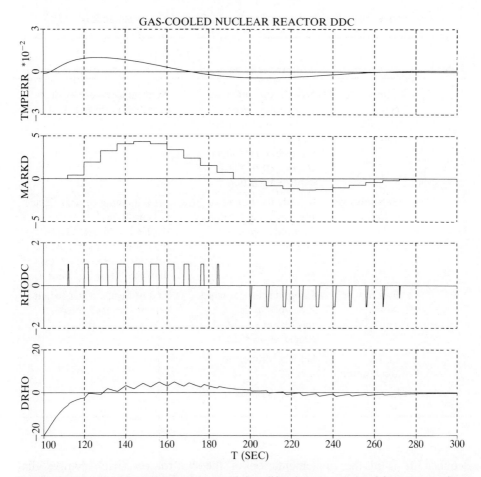

Figure 10.14 Model response to reactivity disturbance: event-driven control system; integrator-driven rod actuation; control action over timescale of 100–300 seconds with output every 0.1 second.

quite good, and the measured gas outlet temperature soon returns to the set point value. Figure 10.14 shows plots for the control actuation for the simulation time range of 100–300 seconds – the region of control activity. The control action is in accord with the temperature error, with no apparent discrepancies to attract our attention.

10.4 IMPROVED SOLUTION USING EVENT PROCESSING

For most of the solution transients obtained from the AGR model, the communication interval forced upon us by the temperature error sampling is quite

Table 10.1 Event processing for AGR model control action

Event number	Description	Process the event	Schedule any other events
1	Temperature error sampling point.	Sample the outlet gas temperature error. Update shift register of past error information. Update the output of the control algorithm.	Next temperature error sampling point (Event 1).
2	Beginning of a 'Space'.	Sample the output of the control algorithm. Compute value of 'Mark'.	(a) Beginning of next 'Space' (Event 2). (b) End of 'Mark' (Event 3). (c) Point at which maximum velocity is reached (if less than the 'Mark') (Event 4).
3	End of 'Mark'.	Compute time at which zero velocity is reached at end of deceleration phase.	End of deceleration (when velocity reaches zero) (Event 5).
4	End of acceleration.	Set acceleration to zero.	
5	End of deceleration.	Set deceleration to zero.	

adequate. It is in the implementation of the control reactivity changes that represent the control actuation, that we need some means of pinpointing the discontinuities that highlight changes in the state of control action. The discrete-event concept of event processing has already come to our rescue for the implementation of discontinuities in Chapter 9. There is no reason why we do not use the same principles to try to overcome what is in reality a very similar problem.

10.4.1 Alterations to the Model

If we refer back to Figure 10.4, we can see that the transients representing control action consist of a series of straight-line segments making up what is therefore a piecewise linear plot. None of the logic expressed in equations (10.29) to (10.36) enables us to determine the exact location, in advance, of any of the discontinuity points shown in detail in Figure 10.4. It is precisely these points that we have to

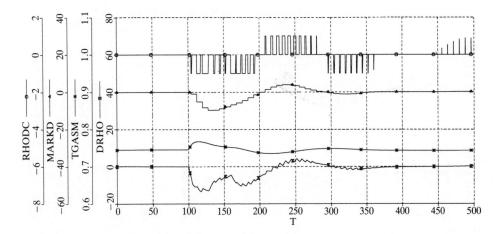

Figure 10.15   Model response to gas flow disturbance: event-driven control system and rod actuation; run-time plot with incorrect actuation performance.

determine in advance in order to control the integration routine to pass over the discontinuities. We also need these points in order to plot the control action transients accurately.

Using the techniques for event-based discrete-event simulation, we may draw up a table to set out the events needed for the model, together with the scheduling and processing arrangements for each.

We have already made a start with two events being scheduled for the implementation of the multirate sampled-data control action. All we are really doing is to supplement these with three further event types to mechanize the reactivity rates of change.

Table 10.1 shows the event processing involved, together with such interaction as is needed between various events. The numbering of events is the same as used in Figure 10.4. The advantage we have in implementing this approach is the fact that all the events may be scheduled in advance. We never 'stumble' upon an unplanned event, and the predictability of every change of direction makes the process very simple. Since the acceleration and deceleration processes are entirely linear, calculation of the various events in the sequence is very easy. The listing for the event-based model is given in the Appendix.

10.4.2   The Gas Flow Disturbance Re-visited

We may proceed to run this new model in a manner similar to the previous one, and we would expect to obtain results very close to those shown in Figures 10.11 and 10.12 for our gas flow disturbance. Figure 10.15 shows the results from the run-time plot. The demanded 'mark' MARKD is identical to that in Figure 10.11,

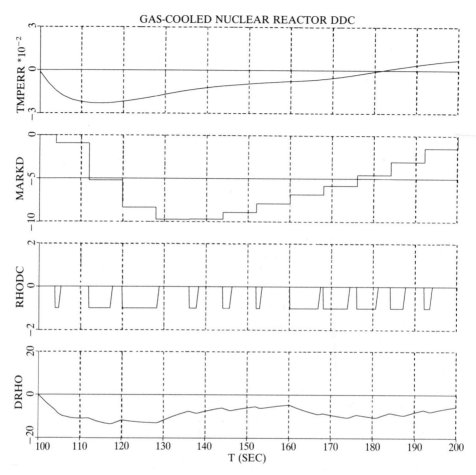

Figure 10.16   Model response to gas flow disturbance: event-driven control system and rod actuation; incorrect control action over timescale of 100–200 seconds.

but the control actuation achieved (RHODC) is somewhat different, as is the overall reactivity change DRHO. It appears that something is wrong with the event processing that vitiates the control actuation to produce erroneous behaviour. Using the same technique as previously, of plotting key variables over a limited portion of the timescale, produces the enlarged transients shown in Figure 10.16. Here we may carry out a semi-quantitative comparison of the demanded 'mark' MARKD with that actually set in by the event processing. Comparison with Figure 10.10 is also useful in this context. For high values of MARKD, instead of 'saturating' to produce constant reactivity decrease at the maximum rate, the control action assumes minimal proportions for instances where MARKD exceeds a value of 8 (the 'space' interval). Clearly this is incorrect, and the cause has to be found.

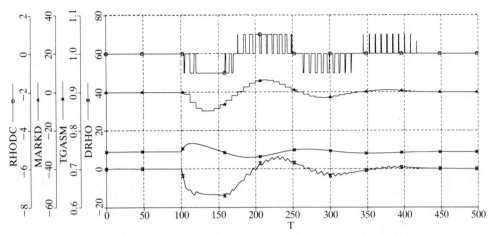

Figure 10.17 Model response to gas flow disturbance: event-driven control system and rod actuation; run-time plot with corrected actuation performance.

Let us look more closely at Figure 10.16 at simulation time $T = 120$ seconds. This marks the beginning of a 'space' for which MARKD has a value of about -8.3. In terms of simulation time, the demanded 'mark' exceeds the 'space' available. On the plot this is reflected by the 'mark' extending beyond the end of the 'space' (simulation time $T = 128$ seconds). At this time, a new 'mark' should be scheduled and the control action continue at maximum velocity. Because the old 'mark' is not cleared, this does not happen and the control action ceases for the rest of the new 'space'. The result is that control action in this region then becomes limited to the excess of 'mark' over 'space' instead of the 'saturation' conditions that are shown in Figure 10.12.

It seems therefore as if the cure for this problem is to disallow scheduling for the end of a 'mark' if that 'mark' exceeds its corresponding 'space'. In that way, control action that is initiated at the beginning of the 'space' will proceed at maximum velocity to the end, and continue at the same pace into the next 'space'. There it will continue to the end or, if the new 'mark' is less than the new 'space', it will terminate in the normal way.

The modification applied to the program is shown in the Appendix, and with this in place, results are obtained as shown in the run-time plot of Figure 10.17. These are almost identical with the transients shown in Figure 10.11, which leads to the conclusion that the simulation is now working correctly. An enlarged plot shown in Figure 10.18 confirms this conclusion when compared with Figure 10.12.

10.4.3 Check on the Reactivity Disturbance

A similar check on the model behaviour after our reactivity disturbance gives rise to transients that are identical to those shown in Figures 10.13 and 10.14. This

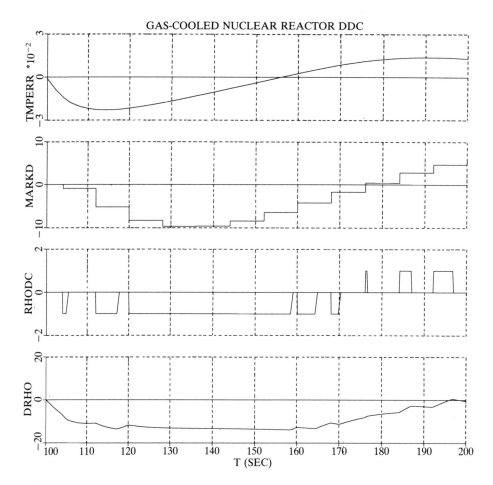

Figure 10.18 Model response to gas flow disturbance: event-driven control system and rod actuation; corrected control action over timescale of 100–200 seconds.

gives confidence that the model is behaving correctly both for conventional and for event-processed control actuation. It does not mean that the model will behave correctly under all circumstances, only that on the basis of the testing and development carried out here, it appears to be doing what we expect it to do.

10.5 CONCLUSIONS

This case study is an example of the development of a model characterized by a great many discontinuities. Besides the expected problems with sampled-data control that are handled by event-processing techniques, the very nature of the control actuation means the introduction of much discontinuous behaviour whenever control actuation takes place.

Conventional programming of the model to implement discontinuous derivative changes at various points in time leads to increased solution times through attempts by the integration routine to negotiate the discontinuity points. What is in many ways worse, however, is the aliasing of the control action transients through lack of output at the discontinuity points. This aliasing is cured by reducing the output interval, but at the price of greatly increased solution times.

By enabling scheduling of the discontinuities, and output at these points in time, event processing assures a faster, more accurate, solution. Solution without event processing of the control actuation can take up to twice as long as the event-processed implementation.

11

Case Study: Simulation of a Distributed-Parameter System

11.1 INTRODUCTION

As we saw earlier, the solution of sets of partial differential equations by means of digital computation techniques is subject not only to the same problems as those encountered for ordinary differential equations, but also to many others, mainly through the introduction of one or more extra independent variables.

In this chapter we shall work through a case study that was based on an actual industrial simulation of a distributed-parameter system, and therefore concerned with the solution of a set of partial differential equations. Although the study was based on a real industrial application, the model developed in this example bears no relation to its industrial counterpart in terms of model parameters and data. Only the solution principles involved are preserved to try to illustrate the techniques used in obtaining a working simulation model.

The production process consists of the two chemical reactions

$A + D \rightarrow B$ (Reaction 1 – endothermic process)

$B + D \rightarrow C + E$ (Reaction 2 – exothermic process)

Component A is a solid which reacts with gas D to produce component B, also a solid. B then reacts with D to produce solid component C and gaseous product E.

One of the objectives of the study was to try to ascertain to what extent the heat of reaction from the second process could be utilized in bringing about the first reaction, without the external addition of heat. For this to be possible, the process would have to be made countercurrent, with the gases flowing in the opposite direction to the solids, and for good utilization of the exothermic heat of reaction by the endothermic process, heat transfer between solid and gas would have to be optimized. This could best be done by having the solid components in fine powder form for maximum heat transfer surface to be available.

The next question that arose concerned the nature of the reaction vessel to be used for a continuous process. Ideally, the powdered solid A would be fed in at one end of a tubular reactor, and the gas D would enter the other end. The problem lay in getting the two phases to mix adequately such that the reactions would proceed stoichiometrically, that is in theoretical proportions of reactants and products everywhere within the reaction vessel. A reactor design was achieved which did meet this requirement to a close degree, which was as well for the simulation exercise, for it enabled considerable simplification of the mathematical model.

In order to be able to start off and sustain the first (endothermic) reaction before the second could get under way, the necessary heat input had to be provided through internal heating coils. Once the second reaction was able to proceed, provision also had to be made for the removal of excess heat through the installation of cooling coils in the reactor vessel. The heating and cooling coils were in close contact with the shell of the vessel; hence vessel shell temperature would be an important consideration in the operation of the simulation model for temperature control purposes.

On the subject of temperature control, we shall soon see that an exothermic reaction can cause considerable control difficulties if there is an upper limit on tolerable reaction temperature. This may be either through safety considerations, or because final product specifications dictate as much. The nature of the temperature control problem may be outlined simply as follows. In a chemical reaction, the rate at which the reaction proceeds (reaction rate) may often be shown to be a function of the reaction temperature according to the relationship

$$k_R = k_0 \exp\left(-\frac{E_A}{RT}\right) \tag{11.1}$$

usually known as the Arrhenius equation in which k_0, E_A and R are constants. The reaction rate k_R will increase in a highly non-linear way with increase in the absolute temperature T at which the reaction proceeds. The exponential increase in reaction rate will then give rise to an increase in the rate of reaction heat produced, which will lead to a further temperature rise. The only way to break this vicious circle is to have some form of temperature control where the excess heat of reaction is removed in order to keep the temperature down.

It was the intention of this simulation study to examine various kinds of control schemes in order to try to meet the powder temperature specifications, and some of these schemes will be discussed in this chapter.

11.2 MODELLING THE PROCESS

The basic partial differential equations used to model the process were based on a general set of equations stated by Himmelblau and Bischoff (1968). These equations were reduced to one spatial dimension in order to represent the dynamics of the tubular reactor used in this process. The general partial differential equations are

$$\frac{\partial c}{\partial t} + \frac{\partial}{\partial z}(v_z c) = \frac{\partial}{\partial z}\left[\rho D_z \cdot \frac{\partial}{\partial z}\left(\frac{c}{\rho}\right)\right] + R \quad \text{(Continuity equation)} \qquad (11.2)$$

$$\rho C_p\left(\frac{\partial T}{\partial t} + v_z \frac{\partial T}{\partial z}\right) = \frac{\partial}{\partial z}\left(K_z \frac{\partial T}{\partial z}\right) + H_R \quad \text{(Energy equation)} \qquad (11.3)$$

where

> c is a term denoting concentration of reacting component
> v_z is velocity of movement through the reactor
> ρ is density of reacting component
> C_p is specific heat
> D_z is a diffusion coefficient in the direction of movement
> R is the rate of reaction
> T is component temperature
> H_R is heat generation through reaction and any radial heat input
> K_z is thermal conductivity
> t is time
> z is distance along the reactor from one end

In deriving a working set of partial differential equations from the general ones above, it is necessary to keep a close watch on the dimensions of the working set, to make sure that all the terms in a particular equation are dimensionally consistent.

Since this example is based on what was a real industrial process, the working set of model equations given below is derived in terms of normalized variables. In this way, the total length of the tubular reaction vessel is given as 1.0, so that intermediate distances in this reactor are always fractional in value. Similarly, all temperatures are normalized with reference to the desired maximum reactor shell temperature hopefully to be achieved by control. The reason for trying to control shell rather than solids temperature was one of available instrumentation. By controlling shell temperature, it was hoped that solids temperature would be kept within specification. A further change imposed upon the general equations stated above is the use of component mole fractions instead of concentrations. This was found to make matters easier in terms of dimension checking.

11.2.1　Working Equations for the Model

For Reaction 1, the rate constant is given by the equation

$$k_{R_1} = 6.5\exp(-2.9/T_S) \qquad (11.4)$$

The reaction rate for this, a second-order reaction, is expressed as

$$R_{R_1} = k_{R_1} \cdot x_A \cdot x_D \tag{11.5}$$

The rate constant for Reaction 2 is given by

$$k_{R_2} = 7.4 \times 10^7 \exp(-10.8/T_S) \tag{11.6}$$

The reaction rate, again second order, is given by the expression

$$R_{R_2} = k_{R_2} \cdot x_B \cdot x_D \tag{11.7}$$

Three continuity equations are required, one for each solids phase component in the system. For components A, B and C respectively, these equations are as follows:

$$\frac{\partial x_A}{\partial t} = 4.57 \times 10^{-4} \frac{\partial^2 x_A}{\partial z^2} - 0.04035 \frac{\partial x_A}{\partial z} - R_{R_1} \tag{11.8}$$

$$\frac{\partial x_B}{\partial t} = 4.57 \times 10^{-4} \frac{\partial^2 x_B}{\partial z^2} - 0.04035 \frac{\partial x_B}{\partial z} + R_{R_1} - R_{R_2} \tag{11.9}$$

$$\frac{\partial x_C}{\partial t} = 4.57 \times 10^{-4} \frac{\partial^2 x_C}{\partial z^2} - 0.04035 \frac{\partial x_C}{\partial z} + R_{R_2} \tag{11.10}$$

where

x_A, x_B, x_C are mole fractions of components A, B and C in the solids
z is the distance along the tubular reaction vessel from the solids inlet end

An energy equation giving the dynamic enthalpy balance over the total solids is written as

$$\frac{\partial T_S}{\partial t} = 4.57 \times 10^{-4} \frac{\partial^2 T_S}{\partial z^2} - 0.04035 \frac{\partial T_S}{\partial z} + 1.235 R_{R_2} - 0.3304 R_{R_1}$$

$$+ 0.2895(T_W - T_S) + 3.9(T_G - T_S) + 0.1995(T_W^4 - T_S^4) \tag{11.11}$$

where

T_S is the temperature of the solids phase
T_W is the tubular reactor shell temperature
T_G is the gas phase temperature

All values are normalized to the reactor shell temperature set point $T_{W_{SP}} = 1.0$. In the above continuity and energy equations, the coefficients of the first derivatives with respect to z have the same value in all the equations. The value represents a velocity of movement through the reactor. The same observation may be made for the coefficients of the second derivatives with respect to z. These values represent a 'back-mixing' effect in the solids passing through the reactor.

As far as the gas-side equations are concerned, a different story emerges. In the general form of the continuity and energy equation, both time and spatial derivatives are present. The dynamics of reaction, mass and heat transfer on the gas side may be shown to have extremely short time constants. This being the case, problems would arise in the numerical integration of such equations resulting in long solution times. It was decided, therefore, to consider the time constants of the gas-side equations to be small enough for time derivative terms to be discarded. The resulting differential equations have spatial derivatives only, as shown below.

As outlined in the introduction, the reactions were known to be nearly stoichiometric everywhere within the reactor. This being the case, simple continuity equations could be written for each of the gas phase components as follows:

$$\frac{\partial x_D}{\partial z} = 24.8(R_{R_1} + R_{R_2}) \tag{11.12}$$

$$\frac{\partial x_E}{\partial z} = -24.8 R_{R_2} \tag{11.13}$$

Similarly, the gas-side enthalpy balance may be represented by means of the following energy equation:

$$(x_D + 1.23 x_E)\frac{\partial T_G}{\partial z} = 888.5(T_G . R_{R_2}) - 3885(T_G . R_{R_1}) - 159(T_S - T_G)$$

$$- 13.2(T_W - T_G) \tag{11.14}$$

Finally, a partial differential equation is required to describe the heat transfer processes that take place in the reactor shell. The shell temperature is defined by the equation

$$\frac{\partial T_W}{\partial t} = 2.93 \times 10^{-5} \frac{\partial^2 T_W}{\partial z^2} + 0.0909(T_S - T_W) + 0.103(T_G - T_W)$$

$$-0.0626(T_W^4 - T_G^4) + \text{Control terms} \tag{11.15}$$

where the control terms are heat transfer contributions, as both heating and cooling in response to temperature control demands. This completes the set of working equations for the model.

11.2.2 Initial and Boundary Conditions

In order to generate a unique solution to the above equations, a set of initial and boundary conditions has to be defined. All the variables are functions of both time and distance through the tubular reactor, and this is taken into account in defining initial and boundary conditions. Take as an example a temperature variable T expressed as $T(z,t)$. At time $t = 0$, T should be defined at all points in space along the whole length of the reactor, to give a proper initial condition or starting point for the solution. The initial condition for an equation solved for T is therefore given by

$$T(z,0) = T_0(z) \quad (0 \leqslant z \leqslant 1.0) \tag{11.16}$$

The spatial domain of solution is the whole length of the reactor vessel. Anything entering that domain will need to have its condition of entry defined over the whole timescale of interest for the simulation, that is for time $t > 0$, as a general case. In this particular problem, the entry of solids material is at one end of the reactor, while the gas flows countercurrent, having entered the other end. Boundary conditions appropriate to the point of entry have to be defined for each component in turn.

For each of the solids phase components, we may write

$$x_A(0,t) = x_{A_0}(t) \quad (t > 0) \tag{11.17}$$

$$x_B(0,t) = x_{B_0}(t) \quad (t > 0) \tag{11.18}$$

$$x_C(0,t) = x_{C_0}(t) \quad (t > 0) \tag{11.19}$$

and for the overall solids phase temperature, we may write

$$T_S(0,t) = T_{S_0}(t) \quad (t > 0) \tag{11.20}$$

all boundary conditions being defined at the solids inlet end of the reactor $(z = 0)$.

For each of the gas phase components, we have

$$x_D(L,t) = x_{D_L}(t) \quad (t > 0) \tag{11.21}$$

$$x_E(L,t) = x_{E_L}(t) \quad (t > 0) \tag{11.22}$$

and for the overall gas temperature

$$T_G(L,t) = T_{G_L}(t) \quad (t > 0) \tag{11.23}$$

where the boundary conditions are defined at the gas inlet end $(z = L = 1.0)$.

For each of those differential equations that are second order with respect to the spatial variable z, it is necessary to define a further boundary condition. This could be similar to those given above, but at the opposite end of the reactor to the inlet. If, as in this case, it is not possible to define appropriate boundary conditions as final values in terms of distance, then some other definition must be found. In the case of the solids phase, it is possible to make reasonable assumptions about the first spatial derivative of each of the relevant solution variables, thereby defining entry conditions for those derivatives.

If it is assumed that prior to entry of the solids phase into the reactor vessel, no reaction, heat transfer or back-mixing of any kind takes place, then we may reasonably assume that the spatial gradient of mole fraction and solids temperature is zero at the point of entry. We may therefore write down the following extra boundary conditions

$$\frac{\partial x_A(0,t)}{\partial z} = 0 \quad (t>0) \tag{11.24}$$

$$\frac{\partial x_B(0,t)}{\partial z} = 0 \quad (t>0) \tag{11.25}$$

$$\frac{\partial x_C(0,t)}{\partial z} = 0 \quad (t>0) \tag{11.26}$$

for the solids phase mole fractions, and for the solids phase temperature, we have

$$\frac{\partial T_S(0,t)}{\partial z} = 0 \quad (t>0) \tag{11.27}$$

One last problem remains with regard to the definition of boundary conditions. The reactor vessel shell temperature equation (11.15) needs initial and boundary conditions similar to those of the variables concerned with the solids phase. It is extremely difficult to define properly a set of boundary conditions that will yield suitable functions of time at each end of the reactor vessel. It is also impossible to make any valid assumptions about the first spatial derivatives of shell temperature at either end of the reactor. In order to arrive at a solution to equation (11.15), therefore, it is necessary to treat the model in such a way that a solution can be obtained. This brings us back to the subject of initial conditions.

Initial conditions need to be defined for each of the partial differential equations in which time derivative terms appear. All except the gas-side equations, therefore, need an initial conditions spatial profile to be defined for the solution variable, in a manner similar to equation (11.16). This is easier said than done for the model in hand. In the simulation that was carried out for the industrial process, the company personnel concerned with the project were anxious to define initial conditions based upon measurement data obtained at many points along the reactor. Although it was possible to arrive at a reasonable-looking profile for each of the system variables, the nature of the process in having two separate reactions taking place consecutively within the same reaction vessel meant that reaction boundaries in particular were hard to quantify. This led to irregular behaviour of the simulated heat transfer processes, and upset the working of the model as a whole. A different tactic had therefore to be employed, one indeed which illustrates the essential steps that should be followed in building up a complex simulation model. The maxim followed was that of starting with the simplest possible model of the system, getting that to work, and then adding and testing further features as required, one at a time, until the whole model was working to everyone's satisfaction. This allowed the adoption of very simple initial conditions profiles for the model. The initial values for each of the solution variables over the whole reactor length were assumed to be equal to the boundary value at inlet. This gave a starting point for the solution of each of the differential equations, which, it was acknowledged, would be quickly superseded by new spatial profiles adopted as the solutions progressed. If we follow the same procedure here, it is first of all necessary to consider how the partial differential equations above might best be turned into the form of first-order ordinary differential equations by the adoption of various finite-difference schemes. The solution can then be handled by a general-purpose digital simulation package for the solution of sets of such equations.

11.2.3 Finite-difference Schemes

Bekey and Karplus (1968) and Schuchmann (1970) have given a lead into the decisions that are necessary to arrive at a suitable finite-difference scheme for the solution of sets of partial differential equations (PDEs) (see Section 9.2). The first possibility that must be considered is that of adopting a wholly finite-element approach to the solution. This discrete-space–discrete-time (DSDT) type of procedure is not really in context with the aims of this book, since it does not involve the solution of ordinary differential equations (ODEs). What we are left with, therefore, is consideration of solution techniques that require the reduction of PDEs to ODEs. In one approach, continuous-space–discrete-time (CSDT) techniques may be employed to remove time derivatives by finite differencing, leaving a set of ODEs to be solved with respect to the spatial variable z. There are several reasons for rejecting this approach. The solution of the resulting ODEs would be a series of spatial profiles whose shapes would change over the course of time, by being updated every finite time step. This would necessitate function storage and playback techniques to handle the profile generation, something that was much favoured in the heyday of the hybrid computer. This would require much memory if a reasonably accurate storage map is to be achieved. The set of ODEs would contain second derivatives with respect to z which would require auxiliary ODEs to be set up for reduction of the system to first order. In this particular problem, doubt was expressed about the ease of solution of the resulting equations, mainly concerning the stability of solution. Finally, in the context of the use to which this particular simulation was to be put, it was felt that a time-transient type of solution would be more in keeping with control tests. Accordingly, it was decided to go for a discrete-space–continuous-time (DSCT) approach, with simple spatial finite differencing to remove spatial derivatives. It was felt that this technique would enable realistic solutions to be generated in both the time and space domains. Time-domain solutions were required for control testing, while spatial profiles were required for model validation. Figure 11.1 illustrates the finite-difference scheme adopted for this problem.

In this diagram, only the system temperatures are shown, but the same reasoning applies to the solids and gas phases with respect to discretization of their respective continuity equations. The finite-element scheme adopted was dictated by the countercurrent nature of the problem. If we consider first the solids temperature, it can be seen that the boundary value T_{S_0} is defined at the solids inlet end, and temperature T_{S_1} can be determined by solving the ODE for solids energy across the first element. The process is repeated across all the remaining elements to define the solution profile for T_S. The process is not as difficult as it might look since all the solids energy ODEs are solved in terms of T_{S_i} ($i = 1, 2, \ldots, n$), which are therefore all state variables.

Discretization of spatial derivatives for the gas-side equations yields sets of algebraic equations for the gas temperature, and for gas-side continuity. These sets of equations have to be solved with care to ensure that they are in the correct computational sequence. The order of solution starts with the boundary conditions at the gas inlet end, and proceeds from right to left in Figure 11.1, in the direction of gas flow.

The one PDE remaining, whose discretization is a problem, is the reactor shell temperature equation. Owing to the radial and axial heat transfer processes that take place across the shell, it is extremely difficult to determine suitable time functions that would serve as boundary conditions at both ends of the reactor. Similarly, it is not easy to specify any spatial derivatives of shell temperature for the same purpose. In order to obtain a solution to this PDE, the discretization process was carried out as shown in Figure 11.1 where the reactor shell temperatures are average values within elements. The way this was made to work will be explained later.

The next stage in the process is to show how each of the terms on the right-hand sides of equations (11.8) to (11.11) are represented in terms of finite differences. Let us first of all consider equation (11.11) with reference to Figure 11.1. The second derivative of T_S with respect to z represents a rate of change of temperature gradient, and in terms of finite differences, we may write

$$\frac{\partial^2 T_S}{\partial z^2} = \frac{\partial}{\partial z}\left(\frac{\partial T_S}{\partial z}\right)$$

$$\cong \frac{1}{\Delta z_i}\left(\frac{\overline{T_{S_{i-1}}} - \overline{T_{S_i}}}{\delta z_{i-1}} - \frac{\overline{T_{S_i}} - \overline{T_{S_{i+1}}}}{\delta z_i}\right) \text{ (for element } i)$$

$$\cong \frac{1}{\Delta z_i}\left(\frac{T_{S_{i-2}} + T_{S_{i-1}} - T_{S_{i-1}} - T_{S_i}}{2\delta z_{i-1}} - \frac{T_{S_{i-1}} + T_{S_i} - T_{S_i} - T_{S_{i+1}}}{2\delta z_i}\right)$$

$$\cong \frac{1}{\Delta z_i}\left(\frac{T_{S_{i-2}} - T_{S_i}}{2\delta z_{i-1}} - \frac{T_{S_{i-1}} - T_{S_{i+1}}}{2\delta z_i}\right) \tag{11.28}$$

In the above equation, the bar over some of the terms indicates an average value for the element in question. At both ends of the reactor, for the first and last elements, the above expression is modified to take account of the boundary

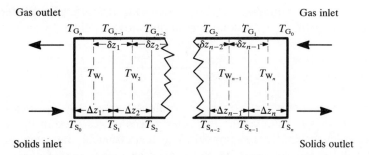

Figure 11.1 Finite-difference scheme for countercurrent reaction process.

condition expressed in equation (11.27). The first spatial derivative is replaced in finite-difference form by the expression

$$\frac{\partial T_S}{\partial z} = \frac{T_{S_i} - T_{S_{i-1}}}{\Delta z_i} \quad \text{(for element } i) \tag{11.29}$$

The remaining terms in equation (11.11) take on average values appropriate to each element in the finite-difference scheme. The average values for reaction rates are derived from the solution of equations (11.4) to (11.7) based on average solids temperature in each element.

It now remains to decide what form the time derivative should have in the ordinary differential equations that replace equation (11.11). Strictly speaking, if one followed the logic of an energy balance over the finite element (from which the general partial differential equation can be derived), then the time derivative of temperature should be based on an average value and a central difference scheme should apply. In view of the problems that this can cause, it was decided in this project to use backward differences instead. The time derivative for the element i was therefore defined in terms of the value at the downstream end of the element, and the equation solved for T_{S_i} $(i = 1, 2, \ldots, n)$.

The same reasoning can be applied to equations (11.8) to (11.10) to obtain similar finite-difference formulations and obtain sets of ODEs to be solved for solids continuity. The same finite differencing is applied to the gas-side equations (11.12) to (11.14), but in this case the results are sets of algebraic equations which have to be solved in strict computational sequence. The one equation that can cause problems in this respect is (11.14) in which T_G appears on both sides of the equation after reduction to algebraic form. This requires some re-arranging in order to obtain a solution for T_G within each element.

It has been mentioned with reference to Figure 11.1 that the shell-side temperature equation (11.15) was treated in a special way for the purposes of finite differencing, in order to be able to solve the resulting ODEs. To avoid the problem of boundary conditions, and to achieve a solution for shell temperature, the average values within elements were set up as the solution variables for the

ODEs derived from equation (11.15). This made things easier in deriving the finite-difference form of the second spatial derivative as given by

$$\frac{\partial^2 T_W}{\partial z^2} = \frac{\partial}{\partial z}\left(\frac{\partial T_W}{\partial z}\right) \cong \frac{1}{\Delta z_i}\left(\frac{\overline{T_{W_{i-1}}} - \overline{T_{W_i}}}{\delta z_{i-1}} - \frac{\overline{T_{W_i}} - \overline{T_{W_{i+1}}}}{\delta z_i}\right) \tag{11.30}$$

for element i where the bars above the elemental values for T_W denote average values, which are the same as the midpoint values represented in Figure 11.1 for which the relevant ODEs are solved. The heat transfer and control terms in equation (11.15) are set up in terms of average values within elements in the resulting ODEs.

11.3 MODEL IMPLEMENTATION

Now that a mathematical model of sorts has been derived, the acid test of its worth is whether or not it can be made to work on a digital computer using a simulation package. This section shows how we may apply the principles outlined in Section 11.2.2, namely starting at the bottom and working up to a more comprehensive model that addresses all required modes of operation.

11.3.1 Radial Heat Transfer

The model that has been described above is required to represent the operation of a tubular reactor in which solids and gas phases, flowing countercurrent, react with one another. The reaction dynamics describe the absorption and generation of heat that take place during the endothermic and exothermic reactions respectively, but during the flow of reactants and products through the reactor, transfer of that heat is going to take place radially through the solids, gas and reactor shell as heat transfer media. Without the reaction processes, the model equations become simply a set of heat transfer equations, and should be tested as such to ensure that the appropriate mechanisms are working correctly. With this in mind, the model was set to operate in the following way.

The boundary conditions for solids and gas temperatures were set to the values

$$T_S(0,t) = T_{S_0}(t) = 0.54 \tag{11.31}$$

$$T_G(L,t) = T_{G_L}(t) = 0.61 \tag{11.32}$$

both values being constant with respect to time. In this example, $L = 1$ is the length of the reactor in normalized units, and the above temperatures are normalized values with respect to the set point for maximum shell temperature control. For simplicity, it was decided to set the initial conditions profile for solids temperature to the boundary value, namely

$$T_S(z,0) = T_{S_0}(z) = 0.54 \tag{11.33}$$

so that in the finite-differenced ODEs, the initial conditions for the elemental solids temperatures were set to the value given in equation (11.33) above. Again, for simplicity, it was decided to set the shell initial conditions temperature profile to be the same as that for the solids, resulting in the relationship

$$T_W(z,0) = T_{W_0}(z) = 0.54 \tag{11.34}$$

In a similar manner, initial conditions profiles were set for the mole fractions of solids reactants and products throughout the length of the reactor as given by

$$x_A(z,0) = x_A(0,t) = 1.0 \tag{11.35}$$

$$x_B(z,0) = x_B(0,t) = 0.0 \tag{11.36}$$

$$x_C(z,0) = x_C(0,t) = 0.0 \tag{11.37}$$

The gas-side equations, being algebraic, did not require the initial conditions profiles to be set. Boundary values only were required to specify inlet conditions in order that the relevant equations could be solved in sequence. For the gas phase components, the boundary conditions used were given by

$$x_D(L,t) = 1.0 \tag{11.38}$$

$$x_E(L,t) = 0.0 \tag{11.39}$$

The above specification of initial and boundary conditions was used for all simulation runs carried out with this model.

It was decided to begin with a model where the reactor length was divided into 20 equal-sized finite elements. If this had proved inadequate, an increase in the number could easily have been arranged. In this initial testing phase, the spatial profiles adopted during the course of a simulation run proved to be all important. Not only was it possible to judge the shape of a particular profile as being up to expectation or not, but by arranging for spatial profiles to be produced at set time intervals during the simulation, it was possible to see whether or not the model reached a steady state. This was important from the point of view of a dynamic check for model stability, in cases where stable operation was to be expected.

Simulation run (#1) was carried out in the absence of both reactions, simply to examine the spatial profiles produced by radial heat transfer alone. Figure 11.2 shows a composite picture of spatial profiles produced every 40 minutes during a simulation run of 200 minutes duration (simulation run time, not elapsed computer time).

It is clear from this figure that there is present a certain amount of what might be termed 'spatial instability'. A steady-state spatial profile due to heat transfer

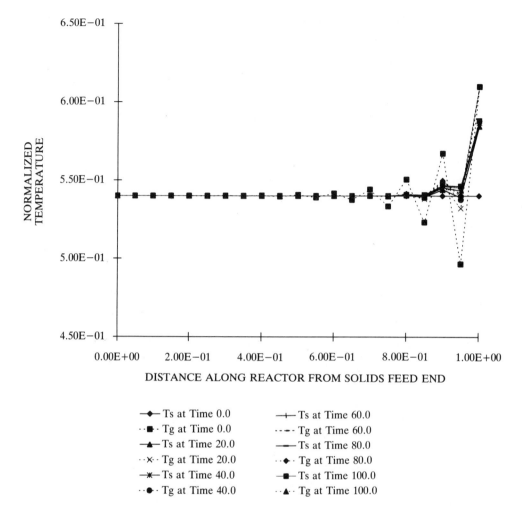

Figure 11.2   Spatial temperature profiles: equal element sizing, no reaction, radial heat transfer only.

effects alone might have been expected to be exponential in shape, somewhat steeply so in this case where heat transfer was known to be efficient. Instead, there appears to be an oscillatory profile, whose amplitude increases towards the gas inlet end. This is reminiscent of the kind of time transient that can be produced by numerical integration routines of fixed step length in the solution of unstable ODEs of exponential order, and therefore suggests that uniform sizing of the spatial finite elements is somehow incorrect for the task in hand.

During the integration of such an ODE using a variable step integration algorithm, as the exponential increases the derivative increases, and the integra-

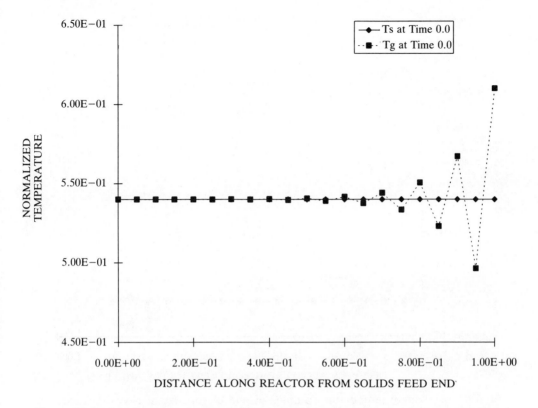

Figure 11.3   Spatial temperature profile at *t* = 0: equal element sizing, no reaction, radial heat transfer only.

tion step length is reduced (commonly by halving). This usually preserves the numerical stability of the solution even if the solution itself depicts the behaviour of an unstable system.

Figure 11.3 shows the spatial profile set produced at time zero. Although the solids profile is at its initial condition value of 0.54, the gas temperature (solved for algebraically) displays the spatial instability discussed above. Figure 11.4 shows the set of two profiles obtained at the end of the simulation run. Something like the rudiments of an exponential shape can be discerned amongst the chaos of the spatial instability.

In a manner analogous to the time-domain solution of an ODE with an increasing exponential solution, it was decided to try to alleviate this spatial instability by the use of differential sizing of the finite elements making up the reactor length. It was obvious, therefore, that the steeper the slope of the temperature profile, the smaller the element size should be. Accordingly, a differential sizing scheme was adopted whereby most of the elements were of a uniform size, but towards the gas inlet end the size was successively halved from

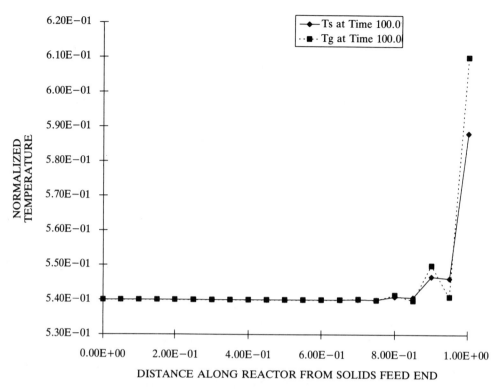

Figure 11.4 Spatial temperature profiles at end of run: equal element sizing, no reaction, radial heat transfer only.

one element to the next over a space of three elements, according to the following rule.

If a length L is divided into N finite elements, a basic finite-element length ΔL can be defined such that the actual lengths that make up the total reactor length are given by

$$\Delta z_i = \Delta L \qquad (i = 1,2,3,\ldots,N-3) \tag{11.40}$$

$$\Delta z_{N-3+j} = \frac{\Delta L}{2^j} \quad (j = 1,2,3) \tag{11.41}$$

In this case this required ΔL to be defined by the relationship

$$\Delta L = \frac{L}{N-2.125} = \frac{1.0}{N-2.125} \quad \text{(for } L = 1.0) \tag{11.42}$$

in order that the sum total of all the elements should add up to L.

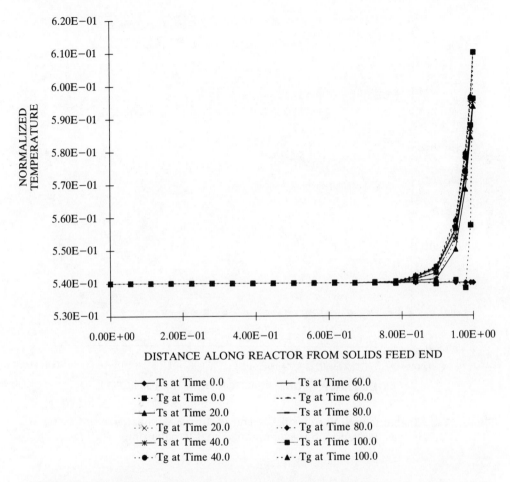

Figure 11.5 Spatial temperature profiles: differential element sizing at gas inlet end, no reaction, radial heat transfer only.

Using the differential sizing described above, run (#2) was carried out under the same conditions as run (#1). Figure 11.5 shows a spatial plot of temperature profiles at 40 minute intervals throughout the simulation run. Apart from the gas temperature profile produced at time zero, which shows a little residual instability close to the gas inlet end, the spatial profiles produced at all other times appear to be reasonably exponential in shape as expected for this heat transfer process. The fact that most of the profiles are bunched together indicates that the heat transfer process reached a steady state very quickly. The results would seem to vindicate the choice of differential element sizing, which was thenceforth adopted as standard for the runs to follow. Figure 11.6 shows the temperature profile set at the end of the simulation run. This confirmed the good approximation to

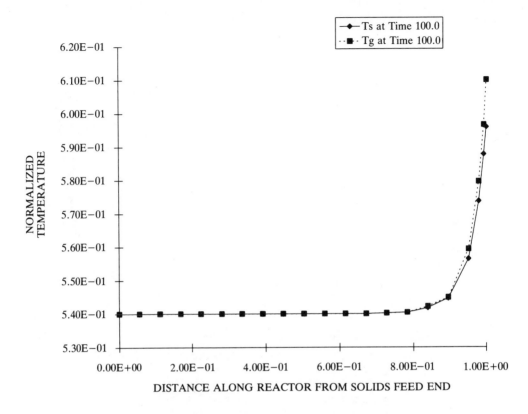

Figure 11.6   Spatial temperature profile at end of run: differential element sizing at gas inlet end, no reaction, radial heat transfer only.

exponential shapes that was obtained by the use of the method. It also confirmed the efficiency of the heat transfer mechanism, on account of both the steep nature of the exponential profiles and the closeness of the solids and gas temperatures. The former indicates that high rates of heat transfer existed, even with the small temperature differences observed between the profiles involved. A qualitative interpretation of the profile shapes of Figure 11.6 is that almost as soon as the gas enters the reactor, it is rapidly cooled down to the inlet temperature of the solids. This observation was considered to be important in anticipating the likely behaviour of the system once the reaction processes were activated.

Although, as shown in Figure 11.6, there is a slightly discontinuous look about the so-called exponential curves generated for the three spatial temperature profiles, it was felt that, at this stage, there was no need to increase the number of elements just to give a smoother appearance to these curves. More finite elements generate more ODEs to be solved, with a consequent increase in solution times.

11.3.2 Introduction of Reaction Dynamics

Once the basic heat transfer mechanisms had been deemed to operate satisfactorily within the model, the next phase of the model testing operation required the addition of reaction dynamics. There were the two reactions to be considered, of which it was anticipated that the exothermic reaction (Reaction 2) could cause most problems in terms of its control. Accordingly, it was decided to introduce to the system the dynamics of Reaction 2, and endeavour to control this reaction on its own.

In order to implement Reaction 2 alone, the initial conditions profiles expressed by equations (11.35) and (11.36) had to be altered. These were re-written as

$$x_A(z,0) = x_A(0,t) = 0.0 \tag{11.43}$$

$$x_B(z,0) = x_B(0,t) = 1.0 \tag{11.44}$$

while all other initial conditions profiles remained as before. This actually made no difference at all to the operation of the powder heat balance, since the heat capacities of all three solids phase components were similar enough for any effect on overall solids heat capacity to be ignored. This assumption was reflected in the construction of equation (11.11). The same was not true for the gas side, as shown by equation (11.14).

The plan of action in introducing Reaction 2 was therefore to assume that the reactor, at the start of the simulation run, was being fed with solids component B only, and that no reaction was taking place, so that B was the only solids component present throughout the length of the reactor. Similarly, gas component D was assumed to be the only constituent of the gas phase throughout the reactor at the start of the run.

This being the state of affairs at run-time zero, Reaction 2 was 'switched on' and the simulation run (#3) allowed to proceed for a run of 100 minutes of simulation time in the total absence of any control, just to observe the behaviour of the system temperatures. Figure 11.7 shows a composite plot of sets of temperature profiles whose 'snapshots' were taken every 20 minutes during the run, starting at time zero.

It was evident that the uncontrolled reaction brought about a steady increase in reaction temperatures, with a maximum in each profile set occurring at about 80% of the reactor length, taken from the solids inlet end. Figure 11.8 shows a composite picture of the corresponding six sets of spatial profiles for mole fractions of solids components B and C. Here it is noticeable that the profiles for 60, 80 and 100 minutes are exactly the same, showing that the composition curves had reached a steady state, even though the temperature was still rising. The steady state is denoted by the termination of the reaction at a point about 84% along the reactor from the solids inlet end. This accounts for the peak reached by

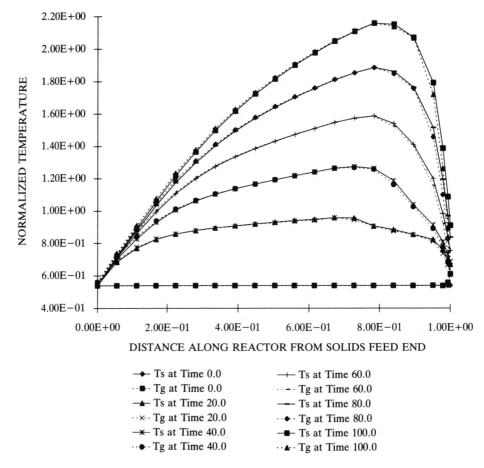

Figure 11.7   Spatial temperature profiles with Reaction 2 activated, no temperature control.

the reaction temperature, which occurred at the point of termination within the reactor.

Although the reactor shell temperature is not shown on the spatial profile plots, it can readily be seen from the simulation run output data that the shell temperature lagged behind the general rise in temperature shown in Figure 11.7. Solids-to-gas heat transfer therefore took place at a far greater rate than that for solids and gas to shell. The cessation of the reaction near the gas inlet end of the reactor brought about a sharp drop in temperatures, as solids were cooled down and gas was heated up. This further vindicated the differential element sizing used at this end of the reactor, for the spatial temperature profiles look smoothly curved in shape.

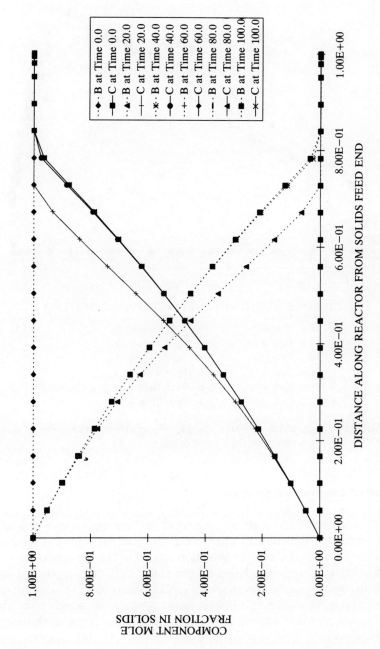

Figure 11.8   Spatial composition profiles during Reaction 2 for components B and C, no temperature control.

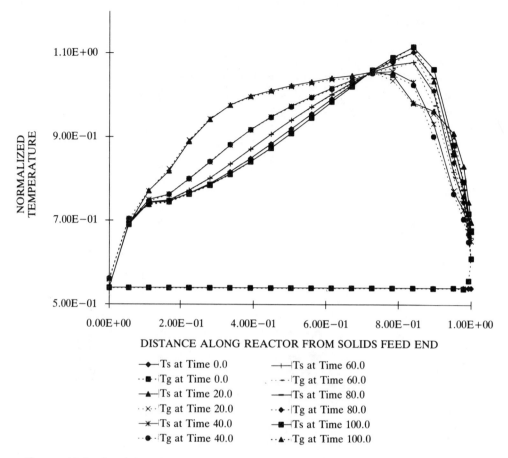

Figure 11.9 Spatial temperature profiles during Reaction 2, with $(P + I)$ point control of shell temperature.

11.3.3 Control of Exothermic Reaction

Observation of the uncontrolled reaction gave some feel for the manner in which the exothermic reaction (Reaction 2) might behave when some kind of control was applied to the system. The control actuation available for the reactor was in the form of heating and cooling applied to the shell, and the purpose of the simulation was to try several modes of control. The real objective was energy conservation by trying to utilize exothermic heat as much as possible for the endothermic reaction, but that control objective is outside the scope of this work and will not be discussed here. From the point of view of this chapter, it is important to see how feasible some form of control is for this system, without spending too much time refining it to an optimum situation.

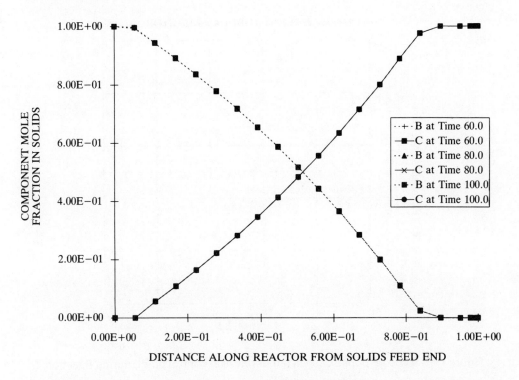

Figure 11.10 Spatial composition profiles towards end of run during Reaction 2, under $(P + I)$ temperature control.

For Reaction 2, a simple proportional-plus-integral $(P + I)$ control system was applied, the temperature measurement being taken at a single point (the point chosen was $z = 0.7$). It was assumed that the heating and cooling of the shell for control actuation was available over half the length of the reactor, at the gas inlet end. This test of Reaction 2 under control constituted run (#4). Figure 11.9 shows a composite picture of the six sets of spatial temperature profiles sampled at intervals of 20 minutes during a total run time of 100 minutes.

Examination of these profiles reveals that the control of shell temperature at the point $z = 0.7$ appeared to be excellent (at that point). Whatever shapes the shell temperature profile might have adopted in the course of time, the point value at the control point remained remarkably constant. Figure 11.10 shows a similar composite set of spatial profiles for the mole fractions of components B and C, taken towards the end of the simulation run at times 60, 80 and 100 minutes. The spatial profile shapes were very similar to those in Figure 11.8, so that the complete set is not shown again. The reaction was complete at more or less the same point ($z = 0.84$), and it is also clear that the temperatures in Figure

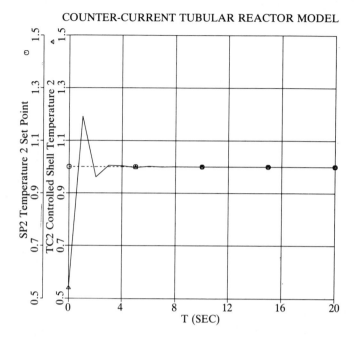

Figure 11.1 Time response of shell temperature at control point, during Reaction 2.

11.9 were trying to increase to form a maximum at about $z = 0.8$, away from the control point. This illustrates the importance of choice for the control point in such a scheme, but the main lesson learned from simulation run (#4) is that the exothermic reaction appeared to be quite controllable, such that progress could be made to the next stage – that of introducing the endothermic reaction – under a similar control scheme. Before doing that, however, one further check had to be made. Figure 11.11 illustrates the time response of the control system, where the set point for control was 1.0 in normalized temperature terms. The control temperature had a settling time of about 5 minutes, with a 12% first peak overshoot, so the control action was near optimal in terms of what it was designed to achieve in the way of point control. This was expected, since shell heating and cooling was used to control shell temperature, not that of solids or gas; hence lags and delays which adversely affected control were kept to a minimum.

11.3.4 Introduction of Endothermic Reaction

After the apparent success in controlling the exothermic reaction, the next step in the development of the simulation model was to add the endothermic reaction.

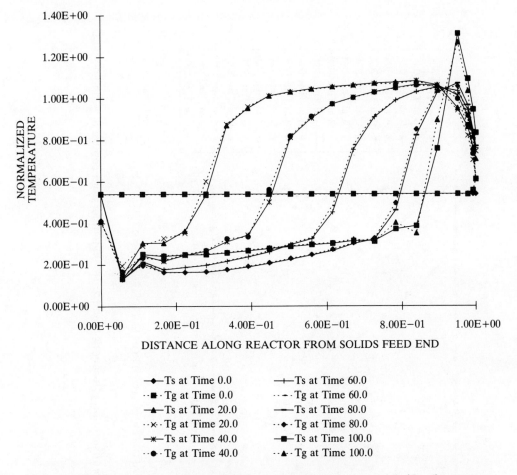

Figure 11.12 Spatial temperature profiles during Reactions 1 and 2: $(P+I)$ temperature control at a single point.

This was quite simply done by altering the initial conditions for the process as
expressed by equations (11.43) and (11.44) to

$$x_A(z,0) = x_A(0,t) = 1.0 \qquad\qquad (11.45)$$

$$x_B(z,0) = x_B(0,t) = 0.0 \qquad\qquad (11.46)$$

which meant that Reaction 1 (the endothermic reaction) was the first to take
place, followed by Reaction 2 once some of component B had been produced.

Figure 11.12 shows a composite set of spatial profiles for solids and gas
temperatures taken every 20 minutes during simulation run (#5) which was also
of 100 minutes duration. At first sight it would seem as if the process did not
reach a steady state during this time, and that the control action was far from

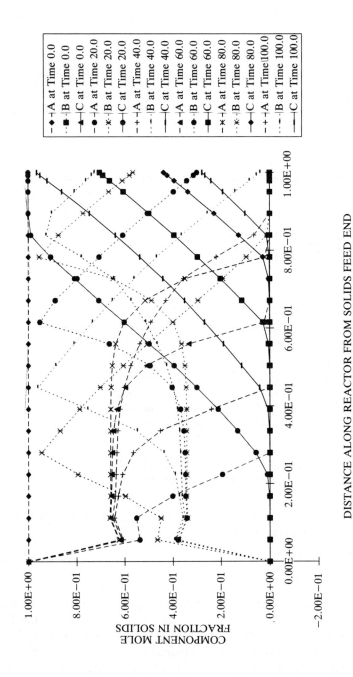

Figure 11.13 Spatial composition profiles during Reactions 1 and 2: $(P+I)$ temperature control at a single point.

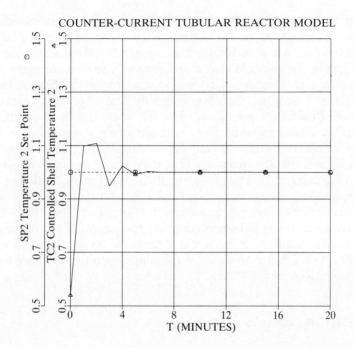

COUNTER-CURRENT TUBULAR REACTOR MODEL

Figure 11.14   Time response of shell temperature at control point during Reactions 1 and 2.

satisfactory. In fact, the model appeared to approach a steady state, as revealed by the 'bunching' of profiles and the formation of a temperature peak at the solids outlet end of the reactor. The same applies to the component mole fractions. In Figure 11.13, we can see that there is a wide spread of spatial profile shapes for the mole fractions of the three solids components present in the system. The following is put forward as a brief interpretation of the results embodied in these two figures.

The temperature control point for run (#5) was chosen to be a distance 90% along the reactor length from the solids inlet end, with control heating and cooling available over the second half of the reactor, at the gas inlet end. It had been estimated that, since the peak temperature for run (#4) occurred at about a distance $z = 0.84$ along the reactor, it would probably occur at a greater distance from the solids inlet end, once Reaction 1 had been brought into play. Accordingly, the temperature control point was placed at a distance $z = 0.9$. Initially when Reaction 1 commenced, the shell temperature was below the set point value, and control heat was input via the jacket to try to raise the shell temperature at the control point ($z = 0.9$) to that value.

Figure 11.14 shows the time transient for the control action at the control point in the reactor, and it is evident that, after a short initial oscillatory transient, the set point was reached very quickly. The initial control heat raised the shell temperature throughout the second half of the reactor to the set point value at the beginning of the run, but once Reaction 1 commenced, there was a temperature drop that began at the solids inlet end and gradually spread along the length of the reactor at a rate roughly equivalent to the transit velocity of the solids. Initially, Reaction 1 went to completion within the first 40% of the reactor length, at which point Reaction 2 got under way. This resulted in Reaction 2 being incomplete at the solids exit point. Since no control heat was supplied to Reaction 1, its rate of reaction slowed, thus moving its completion point further towards the solids exit end of the reactor. This only made matters worse for the completion of Reaction 2. This state of affairs carried on until the reduced temperature reached the control point. An equilibrium of sorts was then approached at which both reactions could take place, but with far from satisfactory results as far as full completion was concerned. If control heating and cooling had been available over the whole reactor, the cooling required to remove the heat of Reaction 2 being produced at the control point would have only made matters worse for Reaction 1.

11.3.5 Control adjustments

It was very clear from the results of simulation run (#5) that temperature control of this process at a single control point in the reactor is quite inadequate. A decision was taken, therefore, to implement two-part temperature control, whereby the reactor was divided into two separate control regions, each with its own control point and temperature set point value. In this way, the reactor was partitioned at $z = 0.2$. The first control zone $(0 \leqslant z \leqslant 0.2)$ was intended to try to ensure that enough control heat was input to drive Reaction 1 to completion within its boundary. Zone 2 $(0.2 \leqslant z \leqslant 1.0)$ was then expected to control Reaction 2 in a manner similar to that displayed in simulation run (#4). Although the Zone 2 set point was still to be 1.0 in normalized units, no particular value stood out as being optimum for Zone 1, so the shell temperature initial condition value of 0.54 in normalized units was chosen. It was hoped that a shell temperature of this magnitude would provide the necessary heat transfer for Reaction 1 to proceed to completion in a short distance along the reactor.

Figure 11.15 shows the now familiar composite set of temperature spatial profiles sampled at 20 minute intervals during simulation run (#6) (duration 100 minutes). The nature of the spatial profiles indicates that the temperature control seemed to work well at the two control points, for at each control point, all solids temperature profiles pass close to the respective set point value, irrespective of the profile shape. The spatial profiles obtained from this run show that Reaction 1 is indeed complete within a short distance into the reactor. Figure 11.16 shows these profiles, revealing that Reaction 2 is complete at a point 10% from the solids outlet end of the reactor, and that the system is in equilibrium with respect

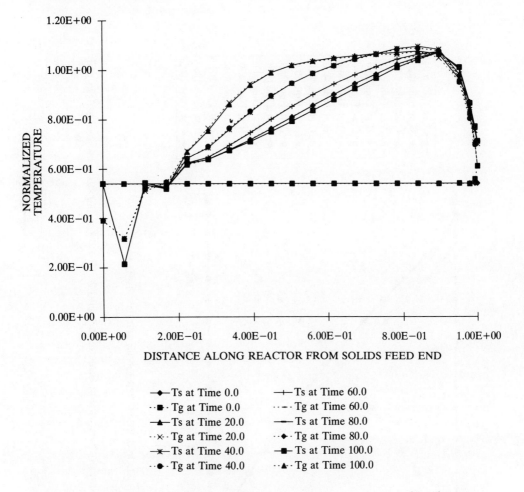

Figure 11.15 Spatial temperature profiles during Reactions 1 and 2: $(P + I)$ temperature control at two points.

to reaction rates, even if not with respect to reaction temperatures. However, one disturbing feature was noticed in regard to the temperatures in Zone 1. There appeared to be some spatial instability present which could be quite easily explained as being due to the discontinuity at the point $z = 0.0$ (solids entry point). The moment the solids entered the reactor, component A came into contact with gas D, Reaction 1 got under way, and the solids temperature dropped sharply. It is obvious that the finite-element spacing at this end of the reactor (which was uniform, unlike the gas inlet end) needed modification. Accordingly, the same differential element sizing was accorded to the solids inlet end of the reactor, as had been implemented in run (#2) at the gas inlet end.

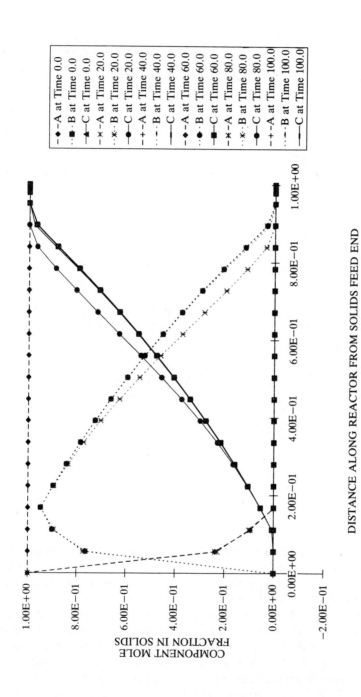

Figure 11.16　Spatial composition profiles during Reactions 1 and 2: $(P + I)$ temperature control at two points.

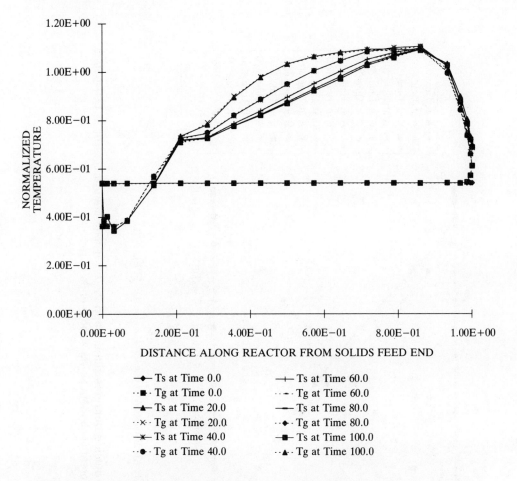

Figure 11.17   Spatial temperature profiles during Reactions 1 and 2: $(P + I)$ temperature control at two points and differential element sizing at both ends of reactor.

Simulation run (#7) of 100 minutes duration was carried out with differential
element sizing in place at both ends of the reactor. Figure 11.17 shows the
composite set of temperature spatial profiles obtained at 20 minute intervals
during the run. It is clear that introduction of the modified finite-element spacing
at the solids inlet end of the reactor had brought about a marked reduction in the
spatial temperature instability featured in the previous simulation run. Although
not altogether absent, there was only a slight degree of spatial oscillation caused
by the temperature discontinuity at the solids entry point. This could have been
reduced further by increasing the number of elements included in the modification
(the counter j in equation (11.41)).

Figure 11.18 shows the composite set of solids component mole fraction

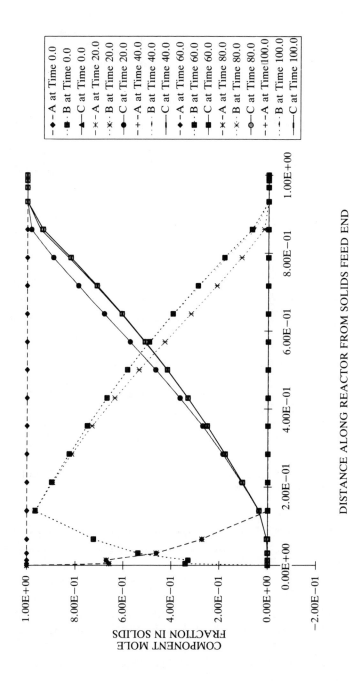

DISTANCE ALONG REACTOR FROM SOLIDS FEED END

Figure 11.18 Spatial composition profiles during Reactions 1 and 2: $(P + l)$ temperature control at two points and differential element sizing at both ends of reactor.

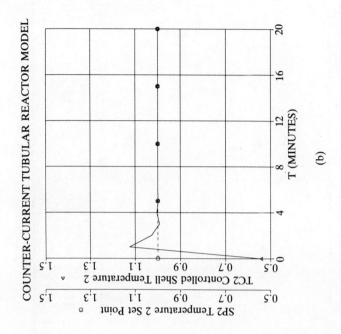

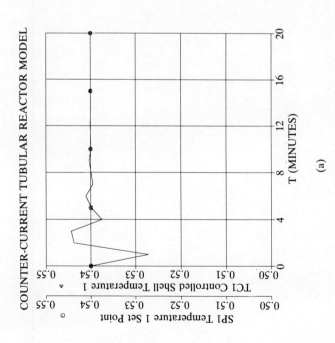

Figure 11.19 Time responses of shell temperature at both control points during Reactions 1 and 2.

Table 11.1   Schedule of run parameters used in the case study

Run number	Figure number	NSPACE	NREACT	NCONT	XAIN	XBIN	MPB	TWCP1	TWCP2	KP1	KI1	KP2	KI2
1	11.2 11.3 11.4	1	0	0	0	1	–	–	–	–	–	–	–
2	11.5 11.6	3	0	0	0	1	–	–	–	–	–	–	–
3	11.7 11.8	3	1	0	0	1	–	–	–	–	–	–	–
4	11.9 11.10 11.11	3	1	1	0	1	0.1	–	0.7	0	0	5.0E−3	0.03
5	11.12 11.13 11.14	3	1	1	1	0	0.1	–	0.9	0	0	5.0E−3	0.03
6	11.15 11.16	3	1	1	1	0	0.2	0.05	0.9	5.0E−3	1.0E−4	5.0E−3	0.03
7	11.17 11.18 11.19	2	1	1	1	0	0.2	0.05	0.9	5.0E−3	1.0E−4	5.0E−3	0.03

spatial profiles for the run. The very sharp decrease in the amount of component A caused a slight spatial oscillation in the profiles for A and B. Apart from this aberration, the profiles seemed to be very well behaved, settling down to a steady state within about 40 minutes of operation. Reaction 1 is complete at about $z = 0.14$ and Reaction 2 at about $z = 0.92$.

Figure 11.19 illustrates the time transients for the separate control actions in Zones 1 and 2 of the reactor. Apart from some initial oscillatory behaviour, the shell temperature control was very good at each control point. With reference to Figure 11.17, it can be seen that the choice of control points was probably reasonable. Zone 1 control point was not considered very critical, but for Zone 2, the value $z = 0.9$ seemed to be a good choice, since the steady state peak temperature reached during the run was close to this point. At this stage, it was decided that this type of control system had been investigated enough to satisfy requirements.

11.4 CONCLUSIONS

This chapter represents an attempt to demonstrate, with respect to a modified industrial system, how a distributed-parameter simulation model is developed and implemented digitally by means of a standard continuous system simulation package for the solution of first-order ordinary differential equations. Much of the intermediate decision-making has been omitted in favour of the presentation of more salient facts, in order not to become too much immersed in detail. It is hoped that the reader will explore for himself or herself some of the detailed examination that is necessary to develop a satisfactory simulation model. The simulation package ACSL/PC for WINDOWS was used to obtain the results discussed in this chapter. Appendix A.1 contains an ACSL listing of the program (files REACTOR.CSL and REACTOR.CMD). Table 11.1 gives a schedule of the runs carried out as described above, together with the model parameter values needed for each run.

11.5 REFERENCES

BEKEY, G A and KARPLUS, W A (1968) *Hybrid computation*, New York: John Wiley, pp. 211–43.

HIMMELBLAU, D M and BISCHOFF, K B (1968) *Process analysis and simulation: deterministic systems*, New York: John Wiley.

SCHUCHMANN, H (1970) 'On the simulation of distributed-parameter systems', *Simulation*, vol. 14, pp. 271–9.

12

Simulation: Which Way Now?

If, in the process of reading all or part of this book, you have found that simulation can indeed be an interesting and fascinating subject, then the main aim of the book will have been achieved. A great deal of satisfaction arises from the accomplishment of creating a simulation model and getting it to work. The hope is that this book will have inculcated into you, the reader, some of the joy (and pain) of simulation model development. If you have been able to try out some of the examples provided, then this process will be more complete.

It now remains for us to take a look into the future for simulation; to try to predict the way forward for simulation techniques. We should also examine some developments that are likely to have considerable impact, both on the ease of interaction with a simulation model and on the scale of modelling problem that may be addressed by simulationists.

12.1 HARDWARE AND SOFTWARE TRENDS

In our brief look at the history of simulation in Section 2.2, we covered that most significant change whereby continuous system models that were at one time implemented only on analog computers, are now almost always programmed into digital processors. We have seen too how there have been significant software developments that have accompanied advances in digital computer hardware. It now remains for us to take a final look at the progress being made in both these areas.

12.1.1 Future Directions in Simulation Hardware Development

Before looking at the way in which computer hardware may develop and its effects upon simulation, it is as well to summarize the areas that have been of greatest benefit up to the present:

- Continuing improvements in raw digital processing speed have allowed faster execution of simulation runs, a consequence of which has been easier development of larger models. User interaction response times have steadily improved for the same reason.

- Ease of access to, and fall in the cost of, computer memory have contributed to the ability to handle large models and vast amounts of data. The same may be said of storage devices such as internal hard-disk drives.

- Improvements in graphics processing and display have enhanced the quality of interaction that a user can have with a simulation model.

- The availability, too, of low-cost high-definition printing enables good presentation of simulation results to be within easy reach. Advances in laser printer technology enable the rapid production of high-quality hard-copy graphics.

- As we have seen, implementation of simulation models within a package such as *ACSL-for-WINDOWS* confers the advantages of a graphics-oriented environment for continuous system simulation. This leads to user interaction at its best.

The enhancements listed above are seen mainly from the viewpoint of the personal computer, but apply equally to minicomputers and workstations. New hardware advances appear almost daily, so there is little point in being very specific about what one might like to see in the future. It is possible, however, to be more general about future hardware advances that will be advantageous in the world of simulation.

For the purposes of simulation, future hardware advances should yield improvements in the following areas:

- Further increases in digital processor speed will always be welcome. At the risk of appearing to overstate the analog computer case, any movement towards the ideal of achieving a very fast turnround time between simulation runs results in a better 'feel' for the model and its behaviour. This is very important for large models.

- Hardware developments that increase user interaction with a simulation model will yield benefits. Some of the interesting developments in 'multimedia' appear to show promise in this direction.

- That nebulous characteristic, 'user-friendliness', can be improved to some extent by hardware developments. Improvements in this area are more likely to arise from enhancements in graphics display technology.

12.1.2 Future Advances in Simulation Software

We have already seen how continuous simulation, as implemented on digital processors, has benefited from the advances in numerical analysis techniques over many years. Without good-quality numerical integration routines, hardware improvements would have relatively little effect, for it is the software implementation that determines the real quality of a simulation exercise. We should therefore give some consideration to those areas of software development that may be good for simulation.

It is true today, and could remain so for some time, that most continuous simulation software packages 'sit on top of' a high-level language compiler. Historically, FORTRAN has been the 'shell' for digital computer implementation and, in view of the huge investment already made, is likely to retain its importance. There is, however, some movement towards the languages of 'C' and 'C++' as compiler 'vehicles', and this may gather momentum in the trend towards the use of object-oriented programming.

From the viewpoint of personal computer implementation, a movement towards 32 bit compilers is a distinct improvement. This confers the advantage of faster processing and data transfer, which is putting digital simulation on these platforms well into the league of real-time working. The portability of personal computers, particularly the 'notebook' or 'laptop' kind, enables digital simulation models to be connected to hardware (e.g. hardware controllers) for real-time testing. There is also the added benefit in 32 bit processing of having access to a larger address space. Any removal of the restrictions of 16 bit computing on model size are bound to find favour. The ability to address an extended memory space puts personal computing on a par with mini- and mainframe machines for the implementation of large simulation problems.

The continued development of graphical user interfaces is a welcome trend. There appears to be a movement towards the adoption of standard protocols for personal computer and workstation use. Simulation software running under these protocols may well have the advantage of running on many different 'platforms', provided that the 'shell' supported by the 'windowing' software insulates sufficiently from the machine hardware. Standardization of operating systems can be of considerable help in this respect.

12.1.3 The Graphical User Interface and Model Definition

Historically, the idea of a user interacting with a computer to define a continuous system model by manipulating icons on a monitor screen is nothing new. In the early days of digital simulation packages, a facility existed for producing a mimic analog computer 'patching' diagram on screen, by moving analog component icons into position and connecting them together to form a model. This model was then translated into a representation that executed with one of the digital simulation packages of the day. The facility was severely limited in scope and size, but was nevertheless a good example of interaction with a graphical user interface for model building.

At the present time, there is a trend towards this kind of operation in the continuing drive towards ever greater user interaction with simulation models. In the area of personal computing, the evident standardization on the WINDOWS graphical user interface has led to the development of interactive facilities for model definition that make good use of that medium. For general-purpose simulation, there are several packages on the market that use the WINDOWS interface. MATLAB 4.0, which was the first version of MATLAB to run under Microsoft WINDOWS, has as an extension, the package SIMULINK, which is a simulation facility whereby models are developed by the manipulation and connection of icons within a window on screen. ACSL, too, has moved in the direction of on-screen model development. The extension, Graphic Modeller, is now becoming available on all hardware platforms that run ACSL, and there is a version of ACSL/GM that runs under Microsoft WINDOWS. For those of us who remember analog computing, the emergence of graphical model building is an exciting prospect, for it once again gives back to the simulationist that intuitive feel that went hand in hand with interactive model development on the analog computer. Having said that, the scope of present-day 'patching diagrams' produced on-screen goes far beyond anything that was available with analog computation. For one thing, an individual icon in SIMULINK can represent a complex concept, such as a time delay function, and, for another, it is possible for the simulationist to construct complex models that can then be represented by a single icon when connected to other equally complex sub-models in a large model structure.

Figure 12.1 shows a composite picture of some of the icon libraries available in SIMULINK. Models may be developed by opening a blank window, 'dragging' icons as required from the library windows and connecting them together to form a model diagram in the new window. Connections may be made to oscilloscope icons, for testing different parts of the model as used to be done in analog computation. Better-quality output plots are obtained through the MATLAB plot facilities.

ACSL Graphic Modeller is a graphics front end for ACSL itself. Figure 12.2 shows a picture of the screen windows that are presented to the user. Again, a comprehensive set of modelling icons is available in each category, and models are put together in much the same way as with SIMULINK.

12.1.4 Example: Simple Glucose Metabolism Model

Figure 12.3 shows a feedback loop block diagram for a model of the human glucose production and metabolism process. This is a simple intuitive model that is derived as follows:

- There is a normal glucose level in the blood G_N which, if increased by an amount G_F by the intake and digestion of food F, registers an error in the glucose level in the blood E_G.
- This causes the pancreas to secrete insulin to a level I_P.
- The insulin level brings about the necessary metabolism to produce a

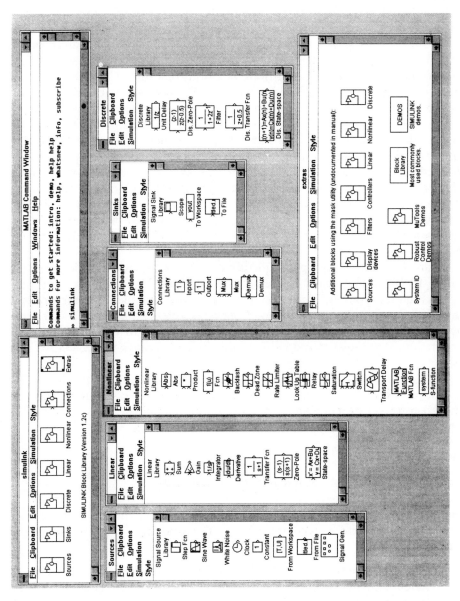

Figure 12.1   Some of the icon libraries in SIMULINK.

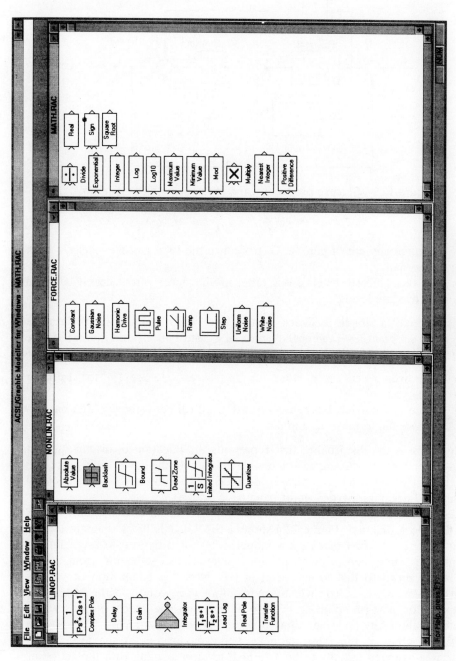

Figure 12.2 ACSL/Graphic Modeller icon groups.

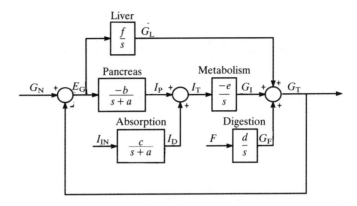

Figure 12.3   Glucose production and metabolism block diagram.

negative amount of glucose G_I to reduce the total glucose level G_T back to normal.

- As the glucose level drops, the insulin level also decays back to steady-state levels.

This is basically a simple feedback control loop which operates to maintain a steady-state set point level under the action of a disturbance (food intake). There is a subsidiary process that operates when blood glucose falls below normal levels:

- The liver serves as a store for any excess glucose (in the form of glycogen).
- When blood levels fall below normal, the liver converts glycogen back to glucose G_L to make up the deficit.

Finally, there is the disturbance that represents the injection of insulin I_{IN} that insulin-dependent diabetics have at intervals throughout the day. This is perceived as an insulin blood level I_D which is assumed to be subject to the same level build-up and decay properties as for pancreatic insulin.

Let us look at a couple of stages of an intuitive model development process using this block diagram. Figure 12.4 shows a SIMULINK block diagram for the testing phase of the first part of this model, where it is desired to ascertain the responses to a unit step decrease in blood glucose level set point, from a high level of 1.0 down to the normal level of 0.0. Note the connection of various outputs to a multiplexer icon which sends output to a scope and the MATLAB workspace for post-run plotting. Figure 12.5 shows the scope output, which indicates qualitatively that everything is operating correctly. Figure 12.6 shows a MATLAB output of the three variables being monitored for the test. Blood glucose was assumed initially to be at a high level in keeping with the set point before the step. Upon triggering of the step, the resulting error initiates insulin

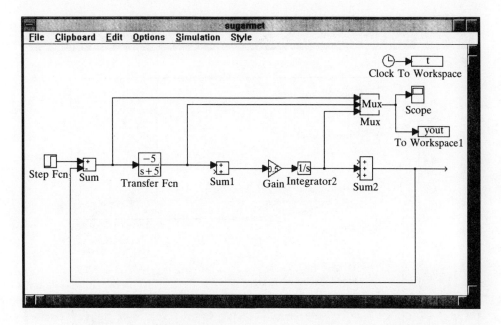

Figure 12.4 SIMULINK diagram for first stage of model testing.

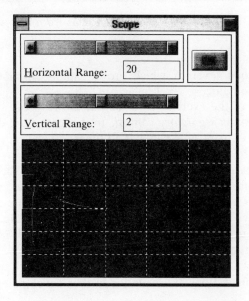

Figure 12.5 'Oscilloscope' output for first-stage testing.

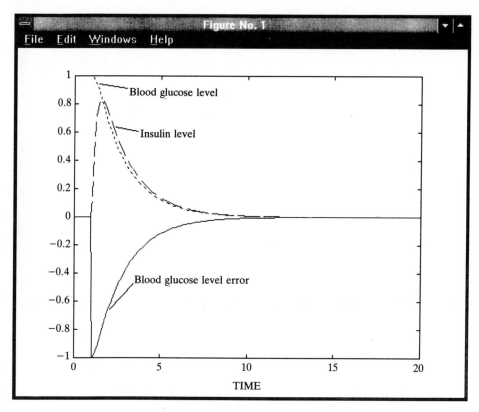

Figure 12.6   MATLAB plot of variables from first-stage tests.

production, the blood level of which grows rapidly at first and then decays naturally as the blood glucose level falls.

The next stage of the exercise is to look at the response of this control loop under what is known as *regulatory control*. In this case, it is desired to keep the blood glucose level at its normal value at all times. However, a disturbance in the form of a rise due to the intake and digestion of food has to be dealt with by a corresponding increase in insulin production to bring about the metabolism of the excess glucose. Figure 12.7 shows the block diagram amended to include the food intake disturbance and the steady-state set point. Note the extra output connections that have been made to monitor the extra variable. Figure 12.8 shows the results plotted in MATLAB. The blood glucose level arising from food is shown as the disturbance. However, the total blood glucose level after insulin control activity shows less of a rise and an eventual fall back to normal.

The following suggestions are made for further development of the model:

- The loop should be added that represents glucose release from the liver.
- The model as shown is entirely linear. There are, however, two non-linear

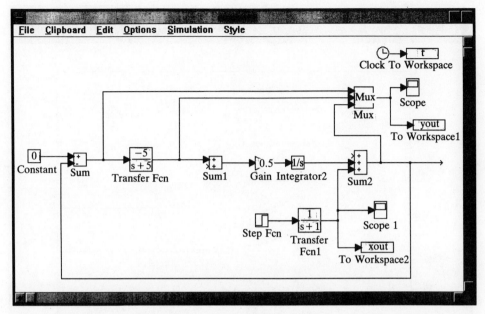

Figure 12.7 SIMULINK diagram for second-stage testing of model response to food intake disturbance.

effects that should be modelled. Normal insulin production is not affected by glucose levels below normal. This is not shown in the linear model and should take the form of some limiting device to prevent negative errors having an effect.

- The liver glucose release would not function if blood glucose levels are above normal. Once again, a limiting device should be added to prevent positive errors affecting the liver glucose release mechanism.
- The normal pancreas operation can be disabled and efforts made to tailor insulin injection disturbances to balance food intake. This is much more difficult than it might seem.

This simple example shows the direction in which general-purpose simulation is progressing. Although it is icon based, graphical modelling is likely to become a standard form of model input. In the advance towards object-oriented programming, the use of icons to represent modelling items of any degree of complexity enhances re-usability and the ability to build up libraries of process items of interest to the particular work in hand.

12.2 PARALLEL PROCESSING

One area of digital computer operation that has been omitted from the above considerations is parallel processing. This deserves to be considered separately,

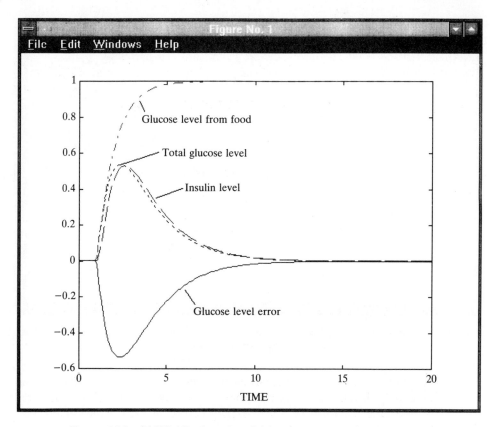

Figure 12.8   MATLAB plot of variables from second-stage tests.

for it is a factor that has implications for both hardware and software development.

When simulation was carried out through the medium of minicomputers and larger machines, one of the most significant enhancements was the arrival of the *array processor*. This is an accessory that provides extra processing power by carrying out the machine arithmetic in a highly specific fashion so as to obtain manifold increases in processing speed. The architecture of the array processor is aimed at providing parallel processing by 'pipelining' and other techniques whereby an arithmetic process is broken down into several steps and passed from one processing element to the next. Many such arithmetic processes take place simultaneously, each being one step ahead of another.

Parallel processing for personal computers and workstations has benefited from the arrival of the *transputer*. This is a device that has the ability to link to many others of its kind in a network of processors that operate truly in parallel. The cheapness of the hardware affords the possibility of achieving large networks of these devices, connected to a personal computer or workstation for interfacing with the outside world.

Much work is being done in the 'transputer' area of parallel processing, and simulation should see the benefit. Specialized languages are used to program 'transputers' and any software developments that enable easier interfacing with these devices from a controlling personal computer or workstation should be encouraged. It is difficult to imagine a limit to the processing power that may be afforded by networks of 'transputers'. There is, however, bound to be much difficulty in harnessing this power through the medium of adequate software. This is needed to control the timing problems involved in synchronizing all the separate operations that together constitute the parallel implementation of a simulation model. Much work remains to be done before parallel processing can become an everyday simulation tool.

12.3 SIMULATION AND 'ARTIFICIAL INTELLIGENCE'

Our look at the future of simulation would not be complete without giving some thought to the application of so-called 'artificial intelligence'. 'Expert system' technology is attracting much interest for possible use in production processes. Fault diagnosis and other control functions are being developed to take advantage of computer technology to employ the decision-making and decision-support facilities afforded by 'expert system' software.

In the field of simulation, there is a similar development trend. Academic researchers have for some time been applying 'expert system' techniques to simulation problems, and clearly work done in this area can pave the way for applications in the 'real' world. In this context, it is perhaps worth remembering that simulation as a tool is used to test out various operating regimes on a model, prior to implementation on a real process. Encouragement should be given to using the same approach in the development of an 'expert system' to be implemented on a real process. Trials conducted with respect to simulation models can help to avoid costly mistakes on the real system.

12.4 EPILOGUE

Simulation has grown from being a set of techniques for modelling a system, which was understood by very few people, to a relatively commonplace industrial activity. The term 'simulation', at one time totally esoteric, has become an everyday word whose superficial meaning, at least, is generally understood. This book is about the techniques of simulation and the way these may be applied in the quest to implement a working model on a digital computer. It is my hope that you have found the book a stimulating introduction to what has been for me a most interesting area of work.

12.5 BIBLIOGRAPHY

MATKO, D, KARBA, R and ZUPANČIČ, B (1992) *Simulation and modelling of continuous systems: A case study approach*, Hemel Hempstead: Prentice Hall.

WILLIAMS, G and PICKUP, J C (Eds.) (1992) *Handbook of diabetes*, Oxford, Blackwell Scientific Publications.

Appendix A.1

ACSL Models Used in this Book

A number of simulation examples in the book were prepared using the ACSL simulation language. The version used was *ACSL for WINDOWS* operating at LEVEL 10 of ACSL. For each simulation example, listings for the model file (*.CSL) and a suggested command file (*.CMD) are given in this appendix. In connection with the command files, the following points should be noted:

- The run execution commands and all plotting commands are formulated into separate procedural blocks so that they may be issued interactively after the program has loaded.
- Run-time plots are produced for the execution of most models.
- All plots result in simulation transients being produced in black. You can introduce colours as you wish.

A.1.1 BEAM.CSL

Copyright © B S Bennett, 1994

```
PROGRAM  BENDING BEAM PROBLEM

    CONSTANT SHEAR0=0.0, BENDM0=0.0, SLOPE0=0.0, DEFL0=0.0, &
             YOUNGM=5.0E09, WEIGHT=100.0, AREAM=2.0E-2, &
             LENGTH=100.0, TOL1=5.0E-4, TOL2=5.0E-3, &
             DELSF=5.0E-5, DELBM=5.0E-4
    VARIABLE X

INITIAL
    1..CONTINUE
END  ! OF INITIAL

DYNAMIC

    DERIVATIVE
        !----------Shear Force on Beam
        SHEAR = INTEG(-WEIGHT/(YOUNGM*AREAM),SHEAR0)
        !----------Bending Moment
```

```
        BENDM = INTEG(SHEAR,BENDM0)
        !----------Slope of Beam
        SLOPE = INTEG(BENDM,SLOPE0)
        !----------Deflection of Beam under its own weight
        DEFL = INTEG(SLOPE,DEFL0)
    END  ! OF DERIVATIVE

    CINT = 0.5
    ALGORITHM IALG = 9
    TERMT(X.GE.LENGTH)

END  ! OF DYNAMIC

TERMINAL
    ARRAY A(2,2), B(2,2), Y(8)
    INTEGER IRUN, ITER
    CONSTANT IRUN=1, ITER=1
    !----------IF FINAL VALUES FOR SLOPE OF DEFLECTION  AND
    !----------DEFLECTION ARE CLOSE ENOUGH TO ZERO, CEASE ITERATION
    IF(ABS(SLOPE).LE.TOL1.AND.ABS(DEFL).LE.TOL2) GO TO 4
    !----------CARRY OUT PERTURBATION OF INITIAL VALUES OF SHEAR FORCE
    !----------AND BENDING MOMENT TO CONSTRUCT SENSITIVITY MATRIX.
        GO TO (11,12,13),ITER


    !----------SAVE FINAL VALUES OF DEFLECTION AND SLOPE
    11..Y(1) = DEFL
        Y(2) = SLOPE
    !----------SAVE INITIAL VALUES OF BENDING MOMENT AND
    !----------SHEAR FORCE (NOT PERTURBED)
        Y(7) = BENDM0
        Y(8) = SHEAR0
    !----------PERTURB INITIAL VALUE OF SHEAR FORCE
      SHEAR0 = Y(8) + DELSF/(10.0**IRUN)
        ITER = ITER + 1
        GO TO 1


    !----------SAVE FINAL VALUES OF DEFLECTION AND SLOPE
    !----------AFTER RUN WITH PERTURBATION OF SHEAR FORCE
    12..Y(3) = DEFL
        Y(4) = SLOPE
    !----------PERTURB INITIAL VALUE OF BENDING MOMENT
      BENDM0 = Y(7) + DELBM/(10.0**IRUN)
    !----------RESET INITIAL VALUE OF SHEAR FORCE TO VALUE
    !----------BEFORE PERTURBATION
      SHEAR0 = Y(8)
        ITER = ITER + 1
        GO TO 1


    !----------SAVE FINAL VALUES OF DEFLECTION AND SLOPE
    !----------AFTER RUN WITH PERTURBATION OF BENDING MOMENT
    13..Y(5) = DEFL
        Y(6) = SLOPE
    !----------CONSTRUCT SENSITIVITY MATRIX
      A(1,1) = (Y(5)-Y(1))/(DELBM/(10.0**IRUN))
      A(1,2) = (Y(3)-Y(1))/(DELSF/(10.0**IRUN))
```

```
      A(2,1) = (Y(6)-Y(2))/(DELBM/(10.0**IRUN))
      A(2,2) = (Y(4)-Y(2))/(DELSF/(10.0**IRUN))
!----------INVERT MATRIX
      CALL MXINV(A,2,B)

!----------CALCULATE NEW INITIAL VALUES FOR BENDING MOMENT
!----------AND SHEAR FORCE
      BENDM0 = Y(7) - B(1,1)*Y(1) - B(1,2)*Y(2)
      SHEAR0 = Y(8) - B(2,1)*Y(1) - B(2,2)*Y(2)
!RESET COUNTER
      ITER = 1
!UPDATE SCALING FACTOR
      IRUN = IRUN + 1
      GO TO 1
    4..CONTINUE
END  ! OF TERMINAL

!----------The FORTRAN subroutine MXINV is listed below
!----------It will be passed straight to the FORTRAN compiler
END  ! OF PROGRAM


!***************************************************************

      SUBROUTINE MXINV(A,N,B)
      DIMENSION A(2,2),B(2,2)
      DIMENSION WORK(10,20)


      DO 1 I=1,N
      DO 1 J=1,N
    1 WORK(I,J) = A(I,J)
      J1 = N+1
      J2 = 2*N
      DO 2 I=1,N
      DO 2 J=J1,J2
    2 WORK(I,J) = 0.0
      DO 3 I=1,N
      J = I+N
    3 WORK(I,J) = 1.0
      DO 610 .K=1,N
      KP1 = K+1
      IF(K.EQ.N) GO TO 500
      L = K
      DO 400 I=KP1,N
  400 IF(ABS(WORK(I,K)).GT.ABS(WORK(L,K))) L = I
      IF(L.EQ.K) GO TO 500
      DO 410 J=K,J2
      TEMP = WORK(K,J)
      WORK(K,J) = WORK(L,J)
  410 WORK(L,J) = TEMP
  500 DO 501 J=KP1,J2
  501 WORK(K,J) = WORK(K,J)/WORK(K,K)
      IF(K.EQ.1) GO TO 600
      KM1 = K-1
      DO 510 I=1,KM1
      DO 510 J=KP1,J2
```

```
510 WORK(I,J) = WORK(I,J) - WORK(I,K)*WORK(K,J)
    IF(K.EQ.N) GO TO 700
600 DO 610 I=KP1,N
    DO 610 J=KP1,J2
610 WORK(I,J) = WORK(I,J) - WORK(I,K)*WORK(K,J)
700 DO 701 I=1,N
    DO 701 J=1,N
    K = J+N
701 B(I,J) = WORK(I,K)
    RETURN
    END
```

A.1.2   BEAM.CMD

Copyright © B S Bennett, 1994

```
SET HVDPRN=.T.
SET FTSPLT=.T.
SET SYMCPL=.F.
SET NPCCPL=20
SET STRPLT=.F.
SET CALPLT=.T.
SET TITLE = 'BENDING BEAM PROBLEM'
OUTPUT X,/LO=0,/HI=100,SHEAR,/LO=-0.0001,/HI=0.0001,BENDM,/LO=-0.001,
&
      /HI=0.001,SLOPE,/LO=-0.01,/HI=0.01,DEFL,/LO=-0.3,/HI=0.1
PREPAR X,SHEAR,BENDM,SLOPE,DEFL
PROCED GO
SPARE
START/PLOT
SPARE
END
PROCED PLOT1
PLOT /XAXIS=X, /XLO=0.0, /XHI=100.0, /XTAG= '(LENGTH)', &
    SHEAR,/COLOR=0,/CHAR=1,BENDM,SLOPE,DEFL
END
PROCED PLOT2
SET SYMCPL=.T.
PLOT /XAXIS=X, /XLO=0.0, /XHI=100.0, /XTAG= '(LENGTH)', &
    SHEAR,/COLOR=0,/CHAR=1,BENDM,SLOPE,DEFL
END
SET CMD=0
```

A.1.3   CONTIN.CSL

Copyright © B S Bennett, 1994

```
PROGRAM SIMPLE CONTINUOUS FEEDBACK SYSTEM

INITIAL
    CONSTANT X10=0.0, X20=0.0, SETPT=1.0, GAIN1=10.0, &
             GAIN2=1.0, POLE=3.5
END  !OF INITIAL
```

```
DYNAMIC
   DERIVATIVE
!----------Compute continuous error
   CERROR = SETPT - X2
!----------Integrator output
   X1 = INTEG(GAIN1*CERROR,X10)
!----------Final output
   X2 = INTEG(X1 - POLE*X2,X20)
   END   !OF DERIVATIVE

   TSTOP = 10.0
   CINT = 0.1
   ALGORITHM IALG=9
   TERMT(T.GE.TSTOP)
END   !OF DYNAMIC

END   !OF PROGRAM
```

A.1.4   CONTIN.CMD

Copyright © B S Bennett, 1994

```
SET STRPLT=.F.
SET CALPLT=.T.
SET SYMCPL=.T.
SET NPCCPL=18
SET TITLE = 'SIMPLE CONTINUOUS FEEDBACK SYSTEM'
OUTPUT /LO=-2,/HI=6,T,/LO=0,/HI=10,X1,/CHAR=1,X2,CERROR
PREPAR T,X1,X2,CERROR
PROCED GO
SPARE
START/PLOT
SPARE
END
PROCED PLOT1
PLOT /XAXIS=T,X1,/COLOR=0,/CHAR=1,X2,CERROR
END
PROCED PLOT2
PLOT /LO=-2.0,/HI=6.0,X1,/COLOR=0,/CHAR=1,X2,CERROR,/SAME
END
PROCED PLOT3
PLOT /XAXIS=X1,/XTAG='(X1)',/XLO=-5.0,/XHI=5.0,/COLOR=0,X2,/CHAR=1
END
PROCED SETSTR
SET CALPLT=.F.
SET STRPLT=.T.
SET SYMSPL=.T.
SET NPCSPL=18
END
PROCED SETCAL
SET STRPLT=.F.
SET CALPLT=.T.
```

```
SET SYMCPL=.T.
SET NPCCPL=18
END
SET CMD=0
```

A.1.5 DDC1.CSL

Copyright © B S Bennett, 1994

```
PROGRAM DDC  ! Normalised gas cooled reactor model
!----------SIMULATION OF DIRECT DIGITAL CONTROL OF THE OUTLET
!----------GAS TEMPERATURE OF AN ADVANCED GAS-COOLED NUCLEAR
!----------POWER STATION REACTOR.

!----------THIS IS AN EXAMPLE OF A MULTI-RATE SAMPLED-DATA
!----------SYSTEM.   THE ERROR IS SAMPLED AND CONTROL ALGORITHM
!----------COMPUTED AT A FREQUENCY WHICH IS A MULTIPLE OF THAT
!----------AT WHICH THE MARK/SPACE CONTROL ROD DRIVE IS ACTUATED.

!----------IN THIS MODEL, THE MARK/SPACE LOGIC OPERATES AS A
!----------CONTINUOUS SYSTEM WITH ITS OWN SET OF INTEGRATORS
!----------DRIVEN BY THE MARK/SPACE CONTROL LOGIC
ARRAY     DHP2(6), DHP20(6), DHP2D(6), A(6), BETA(6), LAMBDA(6), &
          D(6), D0(6), DD(6), BETAD(6), EVENT(5)
CONSTANT  P10=1.02, TFUEL0=1.0, TMOD0=0.5966, TGIN=0.3295, &
          VMAX=1.0,ACCUP=6.6667,ACCDN=10.0,DECUP=5.0,DECDN=1.6667, &
          TC1=1.0, TC2=1.0, TC3=0.05, TC4=20.0, RDIST=-20.0, &
          MGDIST=-0.10, BETAB=0.0055, TGASM0=0.7443, TCLAD0=0.7727, &
          GC=88.0, TS=2.0, SPACE=8.0, RHODC0=0.0, DRHOC0=0.0, &
          TSTOP=500.0, BBMARK=-0.25, TBMARK=0.25, BBTEMP=-0.001, &
          TBTEMP=0.001, TMPREF=0.7443, TRD=1.0E05, TGD=1.0E05
CONSTANT  A = 0.175,0.218,0.184,0.205,0.111,0.107
CONSTANT  BETA = 2.1E-4,1.17E-3,1.04E-3,2.24E-3,7.0E-4,1.4E-4
CONSTANT  LAMBDA = 0.0127,0.0317,0.115,0.311,1.4,3.87
CONSTANT  BETAD = 8.0E-6,5.5E-4,0.00696,0.0417,0.221,0.475
CONSTANT  EVENT = 0.0,0.0,0.0,0.0,0.0

!----------INITIAL REGION
INITIAL
!----------CALCULATION OF INITIAL CONDITIONS FOR DELAYED NEUTRON
!----------AND DECAY HEAT EQUATIONS
      DO 10 J=1,6
      D0(J) = BETA(J) * P10/LAMBDA(J)
      DHP20(J) = A(J)*P10
  10..CONTINUE
!----------CALCULATION OF MARK-SPACE ROD DRIVE PARAMETERS
      RTS2 = 1.0/(TS*TS)
      A1 = (TS+TC1)*(TS+TC2)*RTS2
      A2 = (TS+TC3)*(TS+TC4)*RTS2
      B1 = ((TC1+TC2)*TS+2.0*TC1*TC2)*RTS2
      B2 = ((TC3+TC4)*TS+2.0*TC3*TC4)*RTS2
      C1 = TC1*TC2*RTS2
      C2 = TC3*TC4*RTS2
```

```
      MARKD = 0.0  ! Initialise Control parameters
      UN2=0.0; UN1=0.0; UN=0.0; EN2=0.0; EN1=0.0; EN=0.0
      SCHEDULE SAMPLE .AT. 0.0  ! Initialise control and error
sampling
      SCHEDULE CONTRL .AT. 0.0
END ! OF INITIAL

DYNAMIC
!----------DYNAMIC REGION OF PROGRAM, CONTAINING A DERIVATIVE SECTION
!----------AND FIVE DISCRETE SECTIONS FOR MARK-SPACE LOGIC CONTROL

  DERIVATIVE
    PROCEDURAL
!----------STEP DISTURBANCE IN REACTIVITY
      IF(T.LT.TRD) DRHOD = 0.0
      IF(T.GE.TRD) DRHOD = RDIST
!----------STEP DISTURBANCE IN MASS FLOW OF COOLANT
      IF(T.LT.TGD) MGAS = 1.0
      IF(T.GE.TGD) MGAS = 1.0+MGDIST
!----------DRHOU
      DRHOF = 816.8*(TFUEL0-TFUEL)
!----------DRHOM
      DRHOM = 770.0*(TMOD-TMOD0)
!----------P2
      P2 = DHP2(1)+DHP2(2)+DHP2(3)+DHP2(4)+DHP2(5)+DHP2(6)
!----------REACTOR GAS OUTLET TEMPERATURE ERROR
      TMPERR = TMPREF-TGASM
!----------HF
      HFUEL = 0.8701*P1 + 0.0649*P2
!----------HM
      HMOD = 0.0434*P1 + 0.0216*P2
!----------FUEL TEMPERATURE
      TFUELD = 0.0571*HFUEL - 0.2399*(TFUEL-TCLAD)
      TFUEL = INTEG(TFUELD,TFUEL0)
!----------HEAT TRANSFER COEFFICIENT GAS FLOW TERM
      MGAS8 = MGAS**0.8
!----------TOTAL REACTIVITY CHANGE
      DRHO = DRHOF + DRHOM + DRHOD + DRHOC
!----------SUM OF DELAYED NEUTRON FRACTIONS
      SUMDNF = 0.0
      DO 1 J=1,6
      SUMDNF = SUMDNF + LAMBDA(J)*D(J)
   1..CONTINUE
!----------REACTOR POWER
      P1D = (P1*(DRHO/1.0E05-BETAB) + SUMDNF)/0.001
      P1 = INTEG(P1D,P10)
      DO 2 J=1,6
!----------DELAYED NEUTRON EQUATIONS
      DD(J) = BETA(J)*P1 - LAMBDA(J)*D(J)
!----------DECAY HEAT EQUATIONS
      DHP2D(J) = BETAD(J)*(A(J)*P1 - DHP2(J))
   2..CONTINUE
      D = INTVC(DD,D0)
      DHP2 = INTVC(DHP2D,DHP20)
```

```
!---------AVERAGE GAS TEMPERATURE ACROSS REACTOR
      TGBAR = (0.7161*MGAS8*TCLAD + 0.16577*MGAS8*TMOD + &
           0.8572*MGAS*TGIN)/(0.8572*MGAS + 0.88187*MGAS8)
!---------AVERAGE MODERATOR TEMPERATURE
      TMODD = 0.002344*HMOD - 0.002619*MGAS8*(TMOD-TGBAR)
      TMOD = INTEG(TMODD,TMOD0)
!---------GAS TEMPERATURE AT REACTOR OUTLET
      TGOUT = 2.0*TGBAR - TGIN
!---------MEASURED GAS OUTLET TEMPERATURE
      TGASMD = (TGOUT-TGASM)/(2.5*MGAS**(-0.75))
      TGASM = INTEG(TGASMD,TGASM0)
!---------TC
      TCLADD = 1.8078*(TFUEL-TCLAD) - 1.7466*MGAS8*(TCLAD-TGBAR)
      TCLAD = INTEG(TCLADD,TCLAD0)
!---------CONTROL ROD ACTUATION BY MARK-SPACE DDC
!---------RHODC
      IF(MARKD.LT.0.0) THEN
          IF(T.LT.TFINM) THEN
               IF(RHODC.GT.(-VMAX)) THEN
                   RHODCD=-ACCDN
              ELSE
                   RHODC=-VMAX
                   RHODCD=0.0
              END IF
          ELSE IF(RHODC.LT.0.0) THEN
                   RHODCD=DECDN
              ELSE
                   RHODC=0.0
                   RHODCD=0.0
          END IF
      ELSE
          IF(T.LT.TFINM) THEN
               IF(RHODC.LT.VMAX) THEN
                   RHODCD=ACCUP
              ELSE
                   RHODC=VMAX
                   RHODCD=0.0
              END IF
          ELSE IF(RHODC.GT.0.0) THEN
                   RHODCD=-DECUP
              ELSE
                   RHODC=0.0
                   RHODCD=0.0
          END IF
      END IF


      RHODC = INTEG(RHODCD,RHODC0)
!---------DRHOC
      DRHOCD = RHODC
      DRHOC = INTEG(DRHOCD,DRHOC0)
   END ! OF PROCEDURAL
 END ! OF DERIVATIVE
```

```
  DISCRETE SAMPLE
    PROCEDURAL
!---------COMMUNICATION INTERVAL = TS
!---------ERROR SAMPLING AND CONTROL OUTPUT CALCULATION
      CALL LOGD(.FALSE.)
      UN2 = UN1
      UN1 = UN
      EN2 = EN1
      EN1 = EN
      EN = DEAD(BBTEMP,TBTEMP,TMPERR)
      UN = A1/A2*(EN - B1/A1*EN1 + C1/A1*EN2) + &
          B2/A2*UN1 - C2/A2*UN2
      EVENT(1)=T+TS
      SCHEDULE SAMPLE .AT. EVENT(1)
      CALL LOGD(.FALSE.)
    END ! OF PROCEDURAL
  END ! OF DISCRETE SAMPLE
  DISCRETE CONTRL
!---------COMMUNICATION INTERVAL = SPACE
!---------SAMPLING OF CONTROLLER OUTPUT
    PROCEDURAL
      CALL LOGD(.FALSE.)
      MARK = (GC*UN*SPACE/VMAX)
      MARKD = DEAD(BBMARK,TBMARK,MARK)
      TFINM = EVENT(2)+ABS(MARKD)
      EVENT(2) = T+SPACE
      SCHEDULE CONTRL .AT. EVENT(2)
      CALL LOGD(.FALSE.)
    END ! OF PROCEDURAL
  END ! OF DISCRETE CONTRL

  ALGORITHM IALG=9
  CINTERVAL CINT=20.0
  TERMT(T.GE.TSTOP)

END ! OF DYNAMIC

END ! OF PROGRAM
```

A.1.6   DDC1.CMD

```
SET HVDPRN=.T.
SET ALCPLT=.F.
SET SYMCPL=.T.
SET NPCCPL=80
SET LTFCPL=.T.
SET XINCPL=10
SET PSFCPL=0.85
SET SYMSPL=.T.
SET NPCSPL=40
```

```
     SET XINSPL=10
     SET PSFSPL=0.95
     SET YINSPL=2.0
     SET GLTPLT=200
     SET TITLE = 'GAS-COOLED NUCLEAR REACTOR DDC'
     !OUTPUT T,P1,P2,RHODC,TFUEL,TMOD,TCLAD,TMPERR,TGASM,MARKD,DRHO, &
     !      HFUEL,HMOD,DRHOF,DRHOM,DRHOC,DRHOD,EVENT,/NCIOUT=40
     OUTPUT T,/LO=0,/HI=500,RHODC,/LO=-8,/HI=2,/CHAR=1, &
          MARKD,/LO=-60,/HI=40, &
          TGASM,/LO=0.6,/HI=1.1,/CHAR=11, &
          DRHO,/LO=-20,/HI=80,/CHAR=12
     PREPAR T,P1,P2,RHODC,TFUEL,TMOD,TCLAD,TMPERR,TGASM,MARKD,DRHO, &
          HFUEL,HMOD
     PROCED GO
     SPARE
     START/PLOT
     SPARE
     SET STRPLT=.F.
     SET CALPLT=.T.
     END
     PROCED PLOT1
     PLOT /XAXIS=T, /XTAG='(SEC)', &
          P1,/CHAR=1,/LO=0.95,/HI=1.45, &
          P2,/CHAR=2,/LO=0.95,/HI=1.20, &
          RHODC,/CHAR=11,/LO=-3,/HI=2, &
          TFUEL,/CHAR=12,/LO=0.82,/HI=1.02
     END
     PROCED PLOT1A
     PLOT /XAXIS=T, /XTAG='(SEC)', &
          P1,/CHAR=1, &
          P2,/CHAR=2, &
          RHODC,/CHAR=11 &
          TFUEL,/CHAR=12,
     END
     PROCED PLOT2
     PLOT ,TMOD,/CHAR=1,TCLAD,TMPERR,/CHAR=11,TGASM,/CHAR=12
     END
     PROCED PLOT3
     PLOT MARKD,/CHAR=1,DRHO,HFUEL,/CHAR=11,HMOD,/CHAR=12
     END
     PROCED PLOT4
     PLOT TFUEL,/LO=0.5,/HI=1.3,/CHAR=1,TMOD, &
          TCLAD,/CHAR=11,TGASM,/CHAR=12,/SAME
     END
     PROCED SETSTR
     SET STRPLT=.T.
     SET CALPLT=.F.
     END
     PROCED PLOT5
     PLOT /XAXIS=T, /XTAG='(SEC)',/XLO=100,/XHI=300, &
          /CHAR=' ',DRHO,RHODC,/LO=-2,/HI=2,MARKD,TMPERR
     END
     PROCED PLOT5A
     PLOT /XAXIS=T, /XTAG='(SEC)',/XLO=100,/XHI=200, &
          /CHAR=' ',DRHO,RHODC,/LO=-2,/HI=2,MARKD,TMPERR
     END
```

```
PROCED PLOT6
PLOT /XAXIS=T, /XTAG='(SEC)',/XLO=100,/XHI=160, &
     /CHAR=' ',DRHO,RHODC,/LO=-2,/HI=2,MARKD,TMPERR
END
PROCED PLOT7
PLOT /XAXIS=T, /XTAG='(SEC)',/XLO=160,/XHI=220, &
     /CHAR=' ',DRHO,RHODC,/LO=-2,/HI=2,MARKD,TMPERR
END
PROCED PLOT8
PLOT /XAXIS=T, /XTAG='(SEC)',/XLO=220,/XHI=280, &
     /CHAR=' ',DRHO,RHODC,/LO=-2,/HI=2,MARKD,TMPERR
END
SET CMD=0
```

A.1.7 DDC2.CSL

```
PROGRAM DDC  ! Normalised gas cooled reactor model
!----------SIMULATION OF DIRECT DIGITAL CONTROL OF THE OUTLET
!----------GAS TEMPERATURE OF AN ADVANCED GAS-COOLED NUCLEAR
!----------POWER STATION REACTOR.

!----------THIS IS AN EXAMPLE OF A MULTI-RATE SAMPLED-DATA
!----------SYSTEM.  THE ERROR IS SAMPLED AND CONTROL ALGORITHM
!----------COMPUTED AT A FREQUENCY WHICH IS A MULTIPLE OF THAT
!----------AT WHICH THE MARK/SPACE CONTROL ROD DRIVE IS ACTUATED.

!----------IN THIS MODEL, THE MARK/SPACE LOGIC SETS UP EVENTS WHICH
!----------ENSURE COINCIDENCE OF OPERATING DISCONTINUITIES WITH
!----------THE ENDS OF INTEGRATION STEPS.
ARRAY      DHP2(6), DHP20(6), DHP2D(6), A(6), BETA(6), LAMBDA(6), &
           D(6), D0(6), DD(6), BETAD(6), EVENT(5)
CONSTANT   P10=1.02, TFUEL0=1.0, TMOD0=0.5966, TGIN=0.3295, &
           VMAX=1.0, TAUP=0.15, TADN=0.10, TDUP=0.20, TDDN=0.60, &
           TC1=1.0, TC2=1.0, TC3=0.05, TC4=20.0, RDIST=-20.0, &
           MGDIST=-0.10, BETAB=0.0055, TGASM0=0.7443, TCLAD0=0.7727, &
           GC=88.0, TS=2.0, SPACE=8.0, RHODC0=0.0, DRHOC0=0.0, &
           TSTOP=500.0, BBMARK=-0.25, TBMARK=0.25, BBTEMP=-0.001, &
           TBTEMP=0.001, TMPREF=0.7443, TRD=1.0E05, TGD=1.0E05
CONSTANT   A = 0.175,0.218,0.184,0.205,0.111,0.107
CONSTANT   BETA = 2.1E-4,1.17E-3,1.04E-3,2.24E-3,7.0E-4,1.4E-4
CONSTANT   LAMBDA = 0.0127,0.0317,0.115,0.311,1.4,3.87
CONSTANT   BETAD = 8.0E-6,5.5E-4,0.00696,0.0417,0.221,0.475
CONSTANT   EVENT = 0.0,0.0,0.0,0.0,0.0

!----------INITIAL REGION
INITIAL
!----------CALCULATION OF INITIAL CONDITIONS FOR DELAYED NEUTRON
!----------AND DECAY HEAT EQUATIONS
      DO 10 J=1,6
      D0(J) = BETA(J) * P10/LAMBDA(J)
      DHP20(J) = A(J)*P10
```

```
     10..CONTINUE
!----------CALCULATION OF MARK-SPACE ROD DRIVE PARAMETERS
       VSAUP = VMAX/TAUP
       VSADN = -VMAX/TADN
       VSDUP = -VMAX/TDUP
       VSDDN = VMAX/TDDN
       RTS2 = 1.0/(TS*TS)
       A1 = (TS+TC1)*(TS+TC2)*RTS2
       A2 = (TS+TC3)*(TS+TC4)*RTS2
       B1 = ((TC1+TC2)*TS+2.0*TC1*TC2)*RTS2
       B2 = ((TC3+TC4)*TS+2.0*TC3*TC4)*RTS2
       C1 = TC1*TC2*RTS2
       C2 = TC3*TC4*RTS2
       MARKD = 0.0  ! Initialise Control parameters
       UN2=0.0; UN1=0.0; UN=0.0; EN2=0.0; EN1=0.0; EN=0.0
       SCHEDULE SAMPLE .AT. 0.0  ! Initialise control and error
sampling
       SCHEDULE CONTRL .AT. 0.0
END ! OF INITIAL


DYNAMIC
!----------DYNAMIC REGION OF PROGRAM, CONTAINING A DERIVATIVE SECTION
!----------AND FIVE DISCRETE SECTIONS FOR MARK-SPACE LOGIC CONTROL


  DERIVATIVE
    PROCEDURAL
!----------STEP DISTURBANCE IN REACTIVITY
       IF(T.LT.TRD) DRHOD = 0.0
       IF(T.GE.TRD) DRHOD = RDIST
!----------STEP DISTURBANCE IN MASS FLOW OF COOLANT
       IF(T.LT.TGD) MGAS = 1.0
       IF(T.GE.TGD) MGAS = 1.0+MGDIST
!----------DRHOU
       DRHOF = 816.8*(TFUEL0-TFUEL)
!----------DRHOM
       DRHOM = 770.0*(TMOD-TMOD0)
!----------P2
       P2 = DHP2(1)+DHP2(2)+DHP2(3)+DHP2(4)+DHP2(5)+DHP2(6)
!----------REACTOR GAS OUTLET TEMPERATURE ERROR
       TMPERR = TMPREF-TGASM
!----------HF
       HFUEL = 0.8701*P1 + 0.0649*P2
!----------HM
       HMOD = 0.0434*P1 + 0.0216*P2
!----------FUEL TEMPERATURE
       TFUELD = 0.0571*HFUEL - 0.2399*(TFUEL-TCLAD)
       TFUEL = INTEG(TFUELD,TFUEL0)
!----------HEAT TRANSFER COEFFICIENT GAS FLOW TERM
       MGAS8 = MGAS**0.8
!----------TOTAL REACTIVITY CHANGE
       DRHO = DRHOF + DRHOM + DRHOD + DRHOC
!----------SUM OF DELAYED NEUTRON FRACTIONS
       SUMDNF = 0.0
       DO 1 J=1,6
       SUMDNF = SUMDNF + LAMBDA(J)*D(J)
```

```
   1..CONTINUE
!---------REACTOR POWER
      P1D = (P1*(DRHO/1.0E05-BETAB) + SUMDNF)/0.001
      P1 = INTEG(P1D,P10)
      DO 2 J=1,6
!---------DELAYED NEUTRON EQUATIONS
      DD(J) = BETA(J)*P1 - LAMBDA(J)*D(J)
!---------DECAY HEAT EQUATIONS
      DHP2D(J) = BETAD(J)*(A(J)*P1 - DHP2(J))
   2..CONTINUE
      D = INTVC(DD,D0)
      DHP2 = INTVC(DHP2D,DHP20)
!---------AVERAGE GAS TEMPERATURE ACROSS REACTOR
      TGBAR = (0.7161*MGAS8*TCLAD + 0.16577*MGAS8*TMOD + &
            0.8572*MGAS*TGIN)/(0.8572*MGAS + 0.88187*MGAS8)
!---------AVERAGE MODERATOR TEMPERATURE
      TMODD = 0.002344*HMOD - 0.002619*MGAS8*(TMOD-TGBAR)
      TMOD = INTEG(TMODD,TMOD0)
!---------GAS TEMPERATURE AT REACTOR OUTLET
      TGOUT = 2.0*TGBAR - TGIN
!---------MEASURED GAS OUTLET TEMPERATURE
      TGASMD = (TGOUT-TGASM)/(2.5*MGAS**(-0.75))
      TGASM = INTEG(TGASMD,TGASM0)
!---------TC
      TCLADD = 1.8078*(TFUEL-TCLAD) - 1.7466*MGAS8*(TCLAD-TGBAR)
      TCLAD = INTEG(TCLADD,TCLAD0)
!---------CONTROL ROD ACTUATION BY MARK-SPACE DDC
!---------RHODC
      RHODCD = VDOT
      RHODC = INTEG(RHODCD,RHODC0)
!---------DRHOC
      DRHOCD = RHODC
      DRHOC = INTEG(DRHOCD,DRHOC0)
    END ! OF PROCEDURAL
  END ! OF DERIVATIVE

  DISCRETE SAMPLE
    PROCEDURAL
!---------COMMUNICATION INTERVAL = TS
!---------ERROR SAMPLING AND CONTROL OUTPUT CALCULATION
      CALL LOGD(.FALSE.)
      UN2 = UN1
      UN1 = UN
      EN2 = EN1
      EN1 = EN
      EN = DEAD(BBTEMP,TBTEMP,TMPERR)
      UN = A1/A2*(EN - B1/A1*EN1 + C1/A1*EN2) + &
          B2/A2*UN1 - C2/A2*UN2
      EVENT(1)=T+TS
      SCHEDULE SAMPLE .AT. EVENT(1)
      CALL LOGD(.FALSE.)
    END ! OF PROCEDURAL
  END ! OF DISCRETE SAMPLE
```

```
   DISCRETE CONTRL
!----------COMMUNICATION INTERVAL = SPACE
!----------SAMPLING OF CONTROLLER OUTPUT
     PROCEDURAL
        CALL LOGD(.FALSE.)
        MARK = (GC*UN*SPACE/VMAX)
        MARKD = DEAD(BBMARK,TBMARK,MARK)
        EVENT(3) = T + ABS(MARKD)
        IF(MARKD.GT.0.0.AND.RHODC.LT.0.0) GO TO 22
        IF(MARKD.LT.0.0.AND.RHODC.GT.0.0) GO TO 22
        IF(MARKD.LT.0.0) THEN
          VDOT = VSADN
          EVENT(4) = T + TADN
          IF(RHODC.LE.-VMAX) EVENT(4)=0.0
          IF(RHODC.LE.-VMAX) VDOT=0.0
          IF(RHODC.LE.-VMAX) RHODC=-VMAX
        ELSE IF(MARKD.GT.0.0) THEN
          VDOT = VSAUP
          EVENT(4) = T + TAUP
          IF(RHODC.GE.VMAX) EVENT(4)=0.0
          IF(RHODC.GE.VMAX) VDOT=0.0
          IF(RHODC.GE.VMAX) RHODC=VMAX
        ELSE
     22..VDOT = 0.0
          RHODC = 0.0
          EVENT(4) = 0.0
          EVENT(3) = 0.0
        END IF
        IF(EVENT(3).LE.EVENT(4)) EVENT(4)=0.0
        IF(EVENT(4).NE.0.0) THEN
            SCHEDULE RODMAX .AT. EVENT(4)
        END IF
        SCHEDULE MARKC .AT. EVENT(3)
        EVENT(2) = T+SPACE
        SCHEDULE CONTRL .AT. EVENT(2)
        CALL LOGD(.FALSE.)
     END ! OF PROCEDURAL
   END ! OF DISCRETE CONTRL


   DISCRETE MARKC
!----------EVENT TIME SET BY EVENT(3) - END OF MARK
     PROCEDURAL
        IF(MARKD.LT.0.0.OR.(MARKD.EQ.0.0.AND.RHODC.LT.0.0)) THEN
!----------ROD DECELERATION DOWNWARDS
            VDOT = VSDDN
            EVENT(5) = T - TDDN*RHODC/VMAX
        ELSE IF(MARKD.GT.0.0.OR.(MARKD.EQ.0.0.AND.RHODC.GT.0.0)) THEN
!----------ROD DECELERATION UPWARDS
            VDOT = VSDUP
            EVENT(5) = T + TDUP*RHODC/VMAX
        END IF
        IF(EVENT(5).LT.EVENT(2)) THEN
            SCHEDULE DECEL .AT. EVENT(5)
        END IF
        CALL LOGD(.FALSE.)
```

```
      END ! OF PROCEDURAL
    END ! OF DISCRETE MARKC


  DISCRETE RODMAX
!----------EVENT TIME SET BY EVENT(4) - ROD SATURATION RATE REACHED
      PROCEDURAL
        IF(MARKD.LT.0.0) THEN
            VDOT = 0.0
            RHODC = -VMAX
        ELSE
            VDOT = 0.0
            RHODC = VMAX
        END IF
        EVENT(4) = 0.0
        CALL LOGD(.FALSE.)
      END ! OF PROCEDURAL
    END ! OF DISCRETE RODMAX


  DISCRETE DECEL
!----------EVENT TIME SET BY EVENT(5) - END OF DECELERATION
        VDOT = 0.0
        RHODC = 0.0
        EVENT(5) = 0.0
        CALL LOGD(.FALSE.)
    END ! OF DISCRETE DECEL


  ALGORITHM IALG=9
  CINTERVAL CINT=20.0
    TERMT(T.GE.TSTOP)

  END ! OF DYNAMIC

  END ! OF PROGRAM
```

A.1.8 DDC2.CMD

```
SET HVDPRN=.T.
SET ALCPLT=.F.
SET SYMCPL=.T.
SET NPCCPL=80
SET LTFCPL=.T.
SET XINCPL=10
SET PSFCPL=0.85
SET SYMSPL=.T.
SET NPCSPL=40
SET XINSPL=10
SET PSFSPL=0.95
SET YINSPL=2.0
SET GLTPLT=200
SET TITLE = 'GAS-COOLED NUCLEAR REACTOR DDC'
```

```
!OUTPUT T,P1,P2,RHODC,TFUEL,TMOD,TCLAD,TMPERR,TGASM,MARKD,DRHO, &
!      HFUEL,HMOD,DRHOF,DRHOM,DRHOC,DRHOD,EVENT,/NCIOUT=40
OUTPUT T,/LO=0,/HI=500,RHODC,/LO=-8,/HI=2,/CHAR=1, &
      MARKD,/LO=-60,/HI=40, &
      TGASM,/LO=0.6,/HI=1.1,/CHAR=11, &
      DRHO,/LO=-20,/HI=80,/CHAR=12
PREPAR T,P1,P2,RHODC,TFUEL,TMOD,TCLAD,TMPERR,TGASM,MARKD,DRHO, &
      HFUEL,HMOD
PROCED GO
SPARE
START/PLOT
SPARE
SET STRPLT=.F.
SET CALPLT=.T.
END
PROCED PLOT1
PLOT /XAXIS=T, /XTAG='(SEC)', &
      P1,/CHAR=1,/LO=0.95,/HI=1.45, &
      P2,/CHAR=2,/LO=0.95,/HI=1.20, &
      RHODC,/CHAR=11,/LO=-3,/HI=2, &
      TFUEL,/CHAR=12,/LO=0.82,/HI=1.02
END
PROCED PLOT1A
PLOT /XAXIS=T, /XTAG='(SEC)', &
      P1,/CHAR=1, &
      P2,/CHAR=2, &
      RHODC,/CHAR=11 &
      TFUEL,/CHAR=12,
END
PROCED PLOT2
PLOT ,TMOD,/CHAR=1,TCLAD,TMPERR,/CHAR=11,TGASM,/CHAR=12
END
PROCED PLOT3
PLOT MARKD,/CHAR=1,DRHO,HFUEL,/CHAR=11,HMOD,/CHAR=12
END
PROCED PLOT4
PLOT TFUEL,/LO=0.5,/HI=1.3,/CHAR=1,TMOD, &
      TCLAD,/CHAR=11,TGASM,/CHAR=12,/SAME
END

PROCED SETSTR
SET STRPLT=.T.
SET CALPLT=.F.
END
PROCED PLOT5
PLOT /XAXIS=T, /XTAG='(SEC)',/XLO=100,/XHI=300, &
      /CHAR=' ',MARKD,RHODC,/LO=-2,/HI=2,TMPERR
END
PROCED PLOT6
PLOT /XAXIS=T, /XTAG='(SEC)',/XLO=100,/XHI=160, &
      /CHAR=' ',MARKD,RHODC,/LO=-2,/HI=2,TMPERR
END
```

```
PROCED PLOT7
PLOT /XAXIS=T, /XTAG='(SEC)',/XLO=160,/XHI=220, &
     /CHAR=' ',MARKD,RHODC,/LO=-2,/HI=2,TMPERR
END
PROCED PLOT8
PLOT /XAXIS=T, /XTAG='(SEC)',/XLO=220,/XHI=280, &
     /CHAR=' ',MARKD,RHODC,/LO=-2,/HI=2,TMPERR
END
SET CMD=0
```

A.1.9 HEATBAR.CSL

Copyright © B S Bennett, 1994

```
PROGRAM  FINITE-ELEMENT MODEL OF A HEATED BAR
CONSTANT NEL=20, HEAT=1.0, TSTOP=30, THCOND=50, SPHEAT=5.0, &
         RHO=4.0, LBAR=10.0, NOUT=0, NSPACE=0, NEVENT=1

ARRAY DELZ(20),SDELZ(20),SDZ(20),THETAD(20),THETA(20), &
      THETA0(20),ERR(20)

INTEGER I, NEL, NSPACE, NEVENT

!----------INITIAL REGION
INITIAL
!-----Calculate mesh points for bar length (Length = LBAR)
     IF(NSPACE.EQ.0)THEN
!-----Regular mesh points
        DLBAR = LBAR/NEL
        DO 12 I=1,NEL
        DELZ(I) = DLBAR
   12..CONTINUE
!-----Differential mesh points (smallest at beginning)
     ELSE
        NCOUNT = 0
        DO 14 I=1,NEL
        NCOUNT = NCOUNT + I
   14..CONTINUE
        DLMIN = LBAR/NCOUNT
        DO 15 I=1,NEL
        DELZ(I) = I*DLMIN
   15..CONTINUE
     END IF

     SDELZ(1) = 0.0
     NO = NEL
     DO 19 I=1,NEL
     SDELZ(I+1) = SDELZ(I) + DELZ(I)
  19..CONTINUE
!-----Calculate HGAIN = K/(Cp*Rho)
     HGAIN = THCOND/(SPHEAT*RHO)
     DO 111 I=1,NO
     THETA0(I) = 0.0
```

```
111..CONTINUE
      SCHEDULE SPPROF .AT. 0.0
END   !OF INITIAL


!---------- DYNAMIC REGION
DYNAMIC
  DERIVATIVE
    PROCEDURAL
      DO 28 I=2,NO
      IF(THETA(I).GT.THETA(I-1)) THETA(I)=THETA(I-1)
  28..CONTINUE
      DO 24 I=1,NO
      ERR(I) = 1.0 - THETA(I)
  24..CONTINUE
      THETAD(1) = HGAIN/DELZ(1)*((HEAT-THETA(1))/DELZ(1)- &
        (THETA(1)-THETA(2))/DELZ(2))
      NDO = NO - 1
      DO 22 I=2,NDO
      THETAD(I)=HGAIN/DELZ(I)*((THETA(I-1)-THETA(I))/DELZ(I) - &
        (THETA(I)-THETA(I+1))/DELZ(I+1))
  22..CONTINUE
      THETAD(NO) = HGAIN*(THETA(NO-1) - THETA(NO))/DELZ(NO)

      DO 26 I=1,NO
      IF(ABS(THETAD(I)).LE.1.0E-10)THETAD(I)=0.0
  26..CONTINUE
    END   !OF PROCEDURAL
    THETA = INTVC(THETAD,THETA0)
  END   !OF DERIVATIVE

  DISCRETE SPPROF
!----------OUTPUT SPATIAL PROFILE INFORMATION FOR LATER PLOTTING
    PROCEDURAL
      IF(NEVENT.EQ.1) OPEN (UNIT=11,FILE='SPATIAL.DAT')
      IF(NEVENT.EQ.1) WRITE(11,1000)(SDELZ(I),I=1,NEL+1)
   1000..FORMAT(21E11.4)
      WRITE(11,1001)T,HEAT,(THETA(J),J=1,NEL)
   1001..FORMAT(E11.4,21(' ',E11.4))
      NEVENT = NEVENT + 1
      IF(NOUT.EQ.0) THEN
!-----Event processed plotting of spatial transients
!     (for more equal spacing)
!-----Calculate inter-event time = 1/Sum of derivatives
      SPOUT = 3.0
      ELSE
        SUMTHD = 0.0
        DO 52 I=1,NO
        SUMTHD = SUMTHD + THETAD(I)
   52..CONTINUE
      SPOUT = 1.0/SUMTHD*NEL/7.0
        IF(T+SPOUT.GT.TSTOP) SPOUT = TSTOP - T
      END IF
      SCHEDULE SPPROF .AT. T+SPOUT
    END   !OF PROCEDURAL
  END   !OF DISCRETE SPPROF
```

```
   CINTERVAL CINT = 0.3
   TERMT (T.GE.TSTOP)
   ALGORITHM IALG=9
END  !OF DYNAMIC


TERMINAL
   CLOSE (UNIT=11)
END  !OF TERMINAL


END  !OF PROGRAM
```

A.1.10 HEATBAR.CMD

Copyright © B S Bennett, 1994

```
SET HVDPRN = .T.
SET CALPLT = .T.
SET STRPLT = .F.
SET SYMCPL = .T.
SET NPCCPL = 20
SET TITLE = 'HEATED BAR SIMULATION'
OUTPUT /LO=0, /HI=1.0, T, /HI=30, THETA(5),THETA(10),THETA(15), &
   THETA(20)
PREPAR T,THETA(1),THETA(2),THETA(3),THETA(4),THETA(5),THETA(6), &
   THETA(7),THETA(8),THETA(9),THETA(10),THETA(11),THETA(12), &
   THETA(13),THETA(14),THETA(15),THETA(16),THETA(17),THETA(18), &
   THETA(19),THETA(20)

PROCED GO
   SPARE
   START/PLOT
   SPARE
END
PROCED PLOT1
   PLOT /XAXIS=T, /XTAG='(SEC)', /XLO=0.0, /XHI=30.0, /LO=0.0, &
        /HI=1.0, /COLOR=0,THETA(5),/TAG='(Temp.Sect.5)',/CHAR=1, &
         THETA(10),/TAG='(Temp.Sect.10)', &
         THETA(15),/TAG='(Temp.Sect.15)', &
         THETA(20),/TAG='(Temp.Sect.20)'
END
PROCED PLOT2
   PLOT /XAXIS=T, /XTAG='(SEC)', /XLO=0.0, /XHI=30.0, /LO=0.0, &
      /HI=1.0, /COLOR=0, &
      THETA(1),THETA(2),THETA(3),THETA(4),THETA(5),THETA(6), &
      THETA(7),THETA(8),THETA(9),THETA(10),THETA(11),THETA(12), &
      THETA(13),THETA(14),THETA(15),THETA(16),THETA(17),THETA(18), &
      THETA(19),THETA(20),/SAME, /OVER, /CHAR=1
END
SET CMD = 0
```

A.1.11 MULTR.CSL

Copyright © B S Bennett, 1994

```
PROGRAM  MULTI-RATE SAMPLED-DATA CONTROL SYSTEM

INITIAL
    CONSTANT TS=0.2, SETPT=1.0, VGAIN=2.0, T1=0.5, T2=0.5, T3=0.1, &
             T4=0.1, ULIMIT=15.0, LLIMIT=-15.0, KC=2.0, NT=3.0, &
             VPOS0=0.0, OUTP0=0.0, DOUTP0=0.0, EN2=0.0, EN1=0.0, &
             EN=0.0, UN2=0.0, UN1=0.0, UN=0.0, SERROR=0.0, &
             COUTP=0.0, VRATED=0.0, VRATE=0.0
    !----------Calculate constants for control algorithm
    TS2 = 1.0/(TS*TS)
    A1 = (TS+T3)*(TS+T4)*TS2
    A2 = (TS+T1)*(TS+T2)*TS2
    B1 = ((T3+T4)*TS + 2.0*T3*T4)*TS2
    B2 = ((T1+T2)*TS + 2.0*T1*T2)*TS2
    C1 = T3*T4*TS2
    C2 = T1*T2*TS2
    !---------Initialise discrete sections
        SCHEDULE ERROR  .AT. 0.0
        SCHEDULE CONTRL .AT. 0.0
END  !OF INITIAL

DYNAMIC

    DERIVATIVE
    !----------Continuous error
    CERROR = SETPT - OUTP
    !----------Valve stem position
    VPOS = INTEG(VRATE,VPOS0)
    !----------2nd-order process decomposed into 2 1st order equations
    DOUTP = INTEG(VPOS-4.0*DOUTP-8.0*OUTP,DOUTP0)
    OUTP = INTEG(DOUTP,OUTP0)
    END  !OF DERIVATIVE

    DISCRETE ERROR
        !----------Log all output data before processing the event
        CALL LOGD(.FALSE.)
        !----------The Procedural section below is necessary in order
        !----------to prevent the system interpreting the statements
        !----------therein to be an algebraic loop, when they are
        !----------really nothing of the sort.  The values on the right-
        !----------hand sides are from the previous sampling interval.
        PROCEDURAL (UN=UN,UN1,UN2,EN,EN1,EN2,A1,A2,B1,B2,C1,C2,KC)
        !----------Shift control values down register
        UN2 = UN1
        UN1 = UN
        EN2 = EN1
        EN1 = EN
        !----------Sample continuous error
        SERROR = CERROR
        EN = SERROR
```

```
       !----------Calculate discrete control output
       UN = KC*A2/A1*(EN-B2/A2*EN1+C2/A2*EN2) + B1/A1*UN1 - C1/A1*UN2
       END  !OF PROCEDURAL
          !----------Log all output data after processing the event
          CALL LOGD(.FALSE.)
          !----------Schedule next time for processing event ERROR
          SCHEDULE ERROR .AT. T+TS
       END  !OF DISCRETE ERROR

       DISCRETE CONTRL
          !----------Log all output data before processing event
          CALL LOGD(.FALSE.)
          !----------Sample control output
          COUTP = UN
          !----------Valve rate demand (Control output x Valve gain)
          VRATED = COUTP*VGAIN
          !----------Valve rate limit
          VRATE = BOUND(LLIMIT,ULIMIT,VRATED)
          !----------Log all output data after processing event
          CALL LOGD(.FALSE.)
          !----------Schedule next time for processing event CONTRL
          SCHEDULE CONTRL .AT. T+(TS*NT)
       END  !OF DISCRETE CONTRL

       ALGORITHM IALG = 9
       CINT = 0.1
       TSTOP = 6.0
       TERMT(T.GE.TSTOP)
   END  !OF DYNAMIC

   END  !OF PROGRAM
```

A.1.12 MULTR.CMD

```
SET STRPLT=.F.
SET CALPLT=.T.
SET SYMCPL=.T.
SET NPCCPL=40
SET HVDPRN=.T.
SET TITLE = 'MULTI-RATE SAMPLED-DATA SYSTEM'
OUTPUT /LO=0,/HI=2,T,/HI=6,OUTP
PREPAR T,OUTP,DOUTP,VPOS,CERROR,SERROR,VRATED,VRATE,UN,COUTP
PROCED GO
SPARE
START/PLOT
SPARE
END
PROCED PLOT1
PLOT /XAXIS=T,/XLO=0.0,/XHI=6.0,OUTP,/COLOR=0,/CHAR=1, &
  SERROR,/STYLE=1,CERROR,/STYLE=2
END
```

```
PROCED PLOT2
PLOT /LO=-
5,/HI=25,VPOS,/COLOR=0,/CHAR=1,VRATED,/STYLE=1,VRATE,/STYLE=2
END
PROCED PLOT3
PLOT /LO=-6,/HI=14,COUTP,/COLOR=0,/CHAR=1,UN,/STYLE=1
END
PROCED PLOT4
PLOT CERROR,/COLOR=0,/CHAR=1,SERROR,/STYLE=1
END
SET CMD=0
```

A.1.13 MULTN.CSL

Copyright © B S Bennett, 1994

```
PROGRAM  MULTI-RATE SAMPLED-DATA CONTROL SYSTEM WITH NOISE

INITIAL
    CONSTANT TS=0.2, SETPT=1.0, VGAIN=2.0, T1=0.5, T2=0.5, T3=0.1, &
             T4=0.1, ULIMIT=15.0, LLIMIT=-15.0, KC=2.0, NT=3.0, &
             VPOS0=0.0, OUTP0=0.0, DOUTP0=0.0, EN2=0.0, EN1=0.0, &
             EN=0.0, UN2=0.0, UN1=0.0, UN=0.0, SERROR=0.0, TSTOP=6.0, &
             COUTP=0.0,VRATED=0.0,VRATE=0.0,TNOISE=0.05,NSEED=1234567
    INTEGER NSEED
    !----------Calculate constants for control algorithm
    TS2 = 1.0/(TS*TS)
    A1 = (TS+T3)*(TS+T4)*TS2
    A2 = (TS+T1)*(TS+T2)*TS2
    B1 = ((T3+T4)*TS + 2.0*T3*T4)*TS2
    B2 = ((T1+T2)*TS + 2.0*T1*T2)*TS2
    C1 = T3*T4*TS2
    C2 = T1*T2*TS2
    YRAND = 0.0
    UNIFI(NSEED)
    !---------Initialise discrete sections
        SCHEDULE ERROR   .AT. 0.0
        SCHEDULE CONTRL  .AT. 0.0
        SCHEDULE NOISE   .AT. 0.0
END  !OF INITIAL

DYNAMIC

    DERIVATIVE
    !----------Continuous error
    CERROR = SETPT - OUTP + YRAND
    !----------Valve stem position
    VPOS = INTEG(VRATE,VPOS0)
    !----------2nd-order process decomposed into 2 1st order equations
    DOUTP = INTEG(VPOS-4.0*DOUTP-8.0*OUTP,DOUTP0)
    OUTP = INTEG(DOUTP,OUTP0)
    END  !OF DERIVATIVE
```

```
DISCRETE ERROR
    !----------Log all output data before processing the event
    CALL LOGD(.TRUE.)
    !----------The Procedural section below is necessary in order
    !----------to prevent the system interpreting the statements
    !----------therein to be an algebraic loop, when they are
    !----------really nothing of the sort.  The values on the right-
    !----------hand sides are from the previous sampling interval.
    PROCEDURAL (UN=UN,UN1,UN2,EN,EN1,EN2,A1,A2,B1,B2,C1,C2,KC)
    !----------Shift control values down register
    UN2 = UN1
    UN1 = UN
    EN2 = EN1
    EN1 = EN
    !----------Sample continuous error
    SERROR = CERROR
    EN = SERROR
    !----------Calculate discrete control output
    UN = KC*A2/A1*(EN-B2/A2*EN1+C2/A2*EN2) + B1/A1*UN1 - C1/A1*UN2
    END  !OF PROCEDURAL
    !----------Log all output data after processing the event
    CALL LOGD(.TRUE.)
    !----------Schedule next time for processing event ERROR
    SCHEDULE ERROR .AT. T+TS
END  !OF DISCRETE ERROR


DISCRETE CONTRL
    !----------Log all output data before processing event
    CALL LOGD(.TRUE.)
    !----------Sample control output
    COUTP = UN
    !----------Valve rate demand (Control output X Valve gain)
    VRATED = COUTP*VGAIN
    !----------Valve rate limit
    VRATE = BOUND(LLIMIT,ULIMIT,VRATED)
    !----------Log all output data after processing event
    CALL LOGD(.TRUE.)
    !----------Schedule next time for processing event CONTRL
    SCHEDULE CONTRL .AT. T+(TS*NT)
END   !OF DISCRETE CONTRL


DISCRETE NOISE
    !----------Log all output data before processing noise event
    CALL LOGD(.TRUE.)
    !----------Sample from a uniform distribution with zero mean and
    !          standard deviation 0.02
    YRAND = (UNIF(0,1)-0.5)*0.02*3.4641

    !----------Log all output data after processing noise event
    CALL LOGD(.TRUE.)
    !----------Schedule next time for processing event NOISE
    SCHEDULE NOISE  .AT. T+TNOISE
END   ! OF DISCRETE NOISE
```

```
    ALGORITHM IALG = 9
    CINT = 0.05
    TERMT(T.GE.TSTOP)
END   !OF DYNAMIC

END   !OF PROGRAM
```

A.1.14 MULTN.CMD

Copyright © B S Bennett, 1994

```
SET STRPLT=.F.
SET CALPLT=.T.
SET GRDCPL=.F.
SET SYMCPL=.T.
SET NPCCPL=17
SET HVDPRN=.T.
SET TITLE = 'MULTI-RATE SAMPLED-DATA SYSTEM'
OUTPUT /LO=0,/HI=2,T,/HI=TSTOP,OUTP
PREPAR T,OUTP,DOUTP,VPOS,CERROR,SERROR,VRATED,VRATE,UN,COUTP,YRAND
PROCED GO
SPARE
START/PLOT
SPARE
END
PROCED PLOT1
PLOT /XAXIS=T,/XLO=0.0,/XHI=TSTOP,/LO=-0.2,/HI=1.2,OUTP,/COLOR=0, &
     /CHAR=1,SERROR,/STYLE=1,CERROR,/STYLE=2
END
PROCED PLOT2
PLOT /LO=-
5,/HI=25,VPOS,/COLOR=0,/CHAR=1,VRATED,/STYLE=1,VRATE,/STYLE=2
END
PROCED PLOT3
PLOT /LO=-3,/HI=12,COUTP,/COLOR=0,/CHAR=1,UN,/STYLE=1
END
PROCED PLOT4
PLOT /LO=-0.4,/HI=1.2,CERROR,/COLOR=0,/CHAR=1,SERROR,/STYLE=1, &
     ,YRAND,/LO=-0.2,/HI=1.2/STYLE=2
END
SET CMD=0
```

A.1.15 REACTOR.CSL

Copyright © B S Bennett, 1994

```
PROGRAM  COUNTER-CURRENT TUBULAR REACTOR MODEL

!BOUNDARY CONDITIONS:
!     TSIN   Solids inlet temperature
!     TGIN   Gas inlet temperature
```

```
!       XAIN   Solids inlet mole fraction of component A
!       XBIN   Solids inlet mole fraction of component B
!       XCIN   Solids inlet mole fraction of component C
!       MDIN   Gas component D inlet molar flow rate (multiple of
!                 solids inlet molar flow rate)
!       MEIN   Gas component E inlet molar flow rate (as for MDIN)
!MODEL CONSTANTS AND VARIABLES
!       MPB    'Midpoint' boundary between reactor control zones
!       TWCP1  Zone 1 reactor wall temperature control point
!       TWCP2  Zone 2 reactor wall temperature control point
!       SP1    Zone 1 temperature control set point
!       SP2    Zone 1 temperature control set point
!       AMBT   Ambient temperature (zero point of normalised
!                   temperature scale used in this model)
!       A1,B1  Arrhenius equation constants for reaction 1
!       A2,B2  Arrhenius equation constants for reaction 2
!       RLIMIT Reaction 2 rate limiting term
!       KI1,KI2     Integral control action gains in zones 1 and 2
!       KP1,KP2     Proportional control action gains in zones 1 and 2
!       NEL    Number of finite elements in reactor model
!       IC1,IC2     Integral control terms for zones 1 and 2
!       SPINT  Time interval for sampling spatial profile data
!       TSTOP  Simulation run time
CONSTANT TSIN=0.54, TGIN=0.61, XAIN=0.0, XBIN=1.0, XCIN=0.0,&
         MDIN=2.0, MEIN=0.0, MPB=0.5, TWCP1=0.2, TWCP2=0.7,&
         SP2=1.0, SP1=0.54, AMBT=0.0, A1=1950,&
         B1=2.9, A2=7.4E7, B2=10.8, RLIMIT=0.0693,&
         KI2=0.03, KP2=5.0E-3, KI1=0.03, KP1=5.0E-3,&
         NEL=20, NREACT=0, NSPACE=1, IC10=0.0, IC20=0.0, &
         TSTOP = 100.0, NEVENT=1, SPINT=20.0, NCONT=0
!FINITE ELEMENTS OF MODEL
!       DELZ   Finite element length in model for definition of
!              solids and gas temperatures
!       DZ     Distance between finite element mid-points, for
!              definition of reactor wall temperatures
!       SDELZ  Cumulative sum of finite element lengths DELZ()
!       SDZ    Cumulative sum of mid-point distances DZ()
!MODEL STATE VARIABLES
!       TS     Solids temperature (normalised with AMBT=0.0)
!       TG     Gas temperature (normalised)
!       .TW    Reactor wall temperature (normalised)
!       XA     Mole fraction of component A in solids
!       XB     Mole fraction of component B in solids
!       XC     Mole fraction of component C in solids
!OTHER ALGEBRAIC VARIABLES
!       MD     Molar flow rate of gas component D
!       ME     Molar flow rate of gas component E
!       CONT   Control action exercised within each finite element
!       ENDRC  Endothermic reaction rate constant (reaction 1)
!       ENDRR  Endothermic reaction rate (reaction 1)
!       EXRC   Exothermic reaction rate constant (reaction 2)
!       EXRR   Exothermic reaction rate (Reaction 2)
ARRAY   DELZ(20), DZ(21), SDELZ(21), SDZ(22),&
        TS0(20), TW0(20), XA0(20), XB0(20), XC0(20),&
        TS(20), TW(20), XA(20), XB(20), XC(20), TG(20),&
```

```
          DTS(20), DTW(20), DXA(20), DXB(20), DXC(20),&
          MD(20), ME(20), CONT(20),&
          ENDRC(20), ENDRR(20), EXRC(20), EXRR(20)
!-----The following integers are used as switches in this program:
!-----    NREACT  = 1 (Reaction present)  = 0 (No reaction)
!-----    NSPACE  = 1 (Uniform spacing)   = 2 (Differential spacing)
!-----            = 3 (Differential spacing at gas inlet end only)
!-----    NCONT   = 0 (No control)        = 1 (Control action
enabled)
!-----    CFLAG1 and CFLAG2 are elements within which control points
!          are located for reactions 1 and 2 respectively
!-----    MPFLAG      is the element within which the zone 'mid-point'
!          boundary is located
INTEGER  I, NEL, NSPACE, NREACT, CFLAG1, CFLAG2, MPFLAG, NCONT


INITIAL
!----------INITIALISATION REGION
!-----Definitions for lengths, etc. of elements
      GO TO (10,11,12),NSPACE
!-----Uniform spacing throughout length
   10..DO 110 I=1,NEL
!-----Incremental element sizes (at ends of elements)
      DELZ(I) = 1.0/NEL
  110..CONTINUE
      GO TO 13


!-----Smaller spacing towards both ends
   11..DELX = 1.0/(1.875+NEL-8)
      DO 14 I=1,4
      DELZ(I) = DELX/(2.0**(5-I))
      DELZ(NEL-I+1) = DELZ(I)
   14..CONTINUE
      L4 = NEL-4
      DO 15 I=5,L4
      DELZ(I) = DELX
   15..CONTINUE
      GO TO 13


!-----Smaller spacing towards gas inlet end only
   12..DELX = 1.0/(0.875+NEL-3)
      DO 120 I=1,3
      DELZ(NEL-I+1) = DELX/(2.0**(4-I))
  120..CONTINUE
      L4 = NEL - 3
      DO 121 I=1,L4
      DELZ(I) = DELX
  121..CONTINUE

   13..SDELZ(1) = 0.0
      DO 16 I=1,NEL


!-----Cumulative grid points at ends of elements
      SDELZ(I+1) = SDELZ(I) + DELZ(I)
```

```
!-----Cumulative grid points at element mid points
      SDZ(I) = 0.5*(SDELZ(I)+SDELZ(I+1))
      SDZ(NEL+1) = SDELZ(NEL+1)
  16..CONTINUE
      DZ(1) = SDZ(1)
!-----Calculate wall temperature control point flags
      DO 195 I=1,NEL
      IF(TWCP1.GT.SDZ(I)) GO TO 195
      CFLAG1 = I
      GO TO 196
 195..CONTINUE
 196..DO 197 I=1,NEL
      IF(TWCP2.GT.SDZ(I)) GO TO 197
      CFLAG2 = I
      GO TO 198
 197..CONTINUE
!-----Calculate midpoint boundary flag
 198..DO 19 I=1,NEL
      IF(MPB.GT.SDELZ(I+1)) GO TO 19
      MPFLAG = I
      GO TO 199
  19..CONTINUE
 199..DO 18 I=1,NEL
!-----Incremental element sizes at mid points
      DZ(I+1) = SDZ(I+1) - SDZ(I)
  18..CONTINUE

      DO 17 I=1,NEL
!-----Initial solids temperature
      TS0(I) = TSIN
!-----Initial mole fraction of B
      XB0(I) = XBIN
!-----Initial mole fraction of C
      XC0(I) = XCIN
!-----Initial mole fraction of A
      XA0(I) = XAIN
!-----Initial reactor wall temperature
      TW0(I) = TSIN
!-----Initial flow rate of gas D
      MD(I) = MDIN
!-----Initial flow rate of gas E
      ME(I) = MEIN
!-----Initial gas temperature
      TG(I) = TSIN
  17..CONTINUE
!-----Enable scheduling of events for storing spatial profile data
      SCHEDULE SPPROF .AT. 0.0
      OPEN (UNIT=11,FILE='SPATIAL.DAT')
END  !OF INITIAL


DYNAMIC
!----------DYNAMIC REGION
  DERIVATIVE
  PROCEDURAL
  25..DO 20 I=1,NEL
      IF(TS(I).LT.0.0) TS(I) = 1.0E-09
```

```
!-----Average solids temperature
      IF(I.EQ.1) TSMEAN = 0.5*(TSIN+TS(1))
      IF(I.GT.1) TSMEAN = 0.5*(TS(I-1)+TS(I))
!-----Endothermic Rate Constant
      ENDRC(I) = A1*EXP(-B1/TSMEAN)
      IF(ENDRC(I).LT.0.0) ENDRC(I) = 0.0
!-----Mole fraction of A (average for section)
      IF(I.EQ.1) XAMEAN = 0.5*(XAIN+XA(I))
      IF(I.GT.1) XAMEAN = 0.5*(XA(I-1)+XA(I))
!-----Endothermic Reaction rate (section average)
      IF(XAMEAN.LT.0.0) XAMEAN = 0.0
      ENDRR(I) = ENDRC(I)*XAMEAN
      IF(T.LT.0.1) ENDRR(I) = 0.0
!-----Exothermic Rate constant
      EXRC(I) = A2*EXP(-B2/TSMEAN)
      IF(EXRC(I).LE.1.0E-30) EXRC(I) = 1.0E-30
!-----Mole fraction of D (average for section)
      MDMEAN = 0.5*(MD(I)+MD(I+1))/(MD(I)+ME(I))
      IF(I.EQ.NEL) MDMEAN = 0.5*(MD(I)+MDIN)/(MD(I)+ME(I))
!-----Mole fraction of B (average for section)
      IF(I.EQ.1) XBMEAN = 0.5*(XBIN+XB(I))
      IF(I.GT.1) XBMEAN = 0.5*(XB(I-1)+XB(I))
!-----Exothermic Reaction rate (section average)
      IF(XBMEAN.LT.0.0) XBMEAN=0.0
      IF(MDMEAN.LE.0.0) MDMEAN = 0.0
      IF(XBMEAN.LT.1.0E-30.OR.MDMEAN.LT.1.0E-30) GO TO 26
      EXRR(I) = 3.0*EXRC(I)**0.6667*XBMEAN*MDMEAN
      EXRR(I) = 1.0/(1.0/(RLIMIT*MDMEAN) + 1.0/EXRR(I))
      GO TO 28
   26..EXRR(I) = 0.0
   28..IF(NREACT.EQ.0) EXRR(I) = 0.0
   20..CONTINUE

      DO 21 I=1,NEL
!-----Mass flow rate of D
      IF(I.EQ.1) MD(NEL-I+1) = MDIN - (XC(NEL-I+1)-XC(NEL-I))
      IF(I.GT.1) MD(NEL-I+1) = MD(NEL-I+2) - &
       (XC(NEL-I+1)-XC(NEL-I))
      IF(I.EQ.NEL) MD(NEL-I+1) = MD(NEL-I+2) - (XC(NEL-I+1)-XCIN)
      IF(MD(NEL-I+1).LT.0.0) MD(NEL-I+1) = 0.0
!-----Mass flow rate of E
      IF(I.EQ.1) ME(NEL-I+1) = MEIN + &
       (XC(NEL-I+1)-XC(NEL-I)) - (XA(NEL-I+1)-XA(NEL-I))
      IF(I.GT.1) ME(NEL-I+1) = ME(NEL-I+2) + &
       (XC(NEL-I+1)-XC(NEL-I)) - (XA(NEL-I+1)-XA(NEL-I))
      IF(I.EQ.NEL) ME(NEL-I+1) = ME(NEL-I+2) + &
       (XC(NEL-I+1)-XCIN) - (XA(NEL-I+1)-XAIN)
      IF(ME(NEL-I+1).GT.MDIN) ME(NEL-I+1) = MDIN
!-----GAS HEAT BALANCE
!-----Average mass flow of D
      IF(I.EQ.1) MDMEAN = 0.5*(MD(NEL-I+1)+MDIN)
      IF(I.GT.1) MDMEAN = 0.5*(MD(NEL-I+1)+MD(NEL-I+2))
!-----Average mass flow of E
      IF(I.EQ.1) MEMEAN = 0.5*(ME(NEL-I+1)+MEIN)
      IF(I.GT.1) MEMEAN = 0.5*(ME(NEL-I+1)+ME(NEL-I+2))
```

```
!-----Average solids temperature
      IF(I.EQ.NEL) TSMEAN = 0.5*(TSIN+TS(1))
      IF(I.LT.NEL) TSMEAN = 0.5*(TS(NEL-I)+TS(NEL-I+1))
!-----Gas temperature
      AA = 29.3*MDMEAN + 36.0*MEMEAN
      BB = (86.78*EXRR(NEL-I+1) + 466.27*ENDRR(NEL-I+1) + &
        5044.865)*DELZ(NEL-I+1)
      CC = (9313.6*TSMEAN + 776.13*TW(NEL-I+1))*DELZ(NEL-I+1)
      AAA = AA + BB
      BBB = AA - BB
      IF(I.EQ.1) TG(NEL-I+1) = (BBB*TGIN+CC)/AAA
      IF(I.GT.1) TG(NEL-I+1) = (BBB*TG(NEL-I+2) + CC)/AAA
   21..CONTINUE


!-----REACTOR WALL TEMPERATURE CONTROL
!-----Interpolate reaction vessel wall temperature at control points
      TC2 = TW(CFLAG2-1) + (TWCP2-SDZ(CFLAG2-1))* &
        (TW(CFLAG2)-TW(CFLAG2-1))/(SDZ(CFLAG2)-SDZ(CFLAG2-1))
      TC1 = TW(CFLAG1-1) + (TWCP1-SDZ(CFLAG1-1))* &
        (TW(CFLAG1)-TW(CFLAG1-1))/(SDZ(CFLAG1)-SDZ(CFLAG1-1))
!-----Temperature error
      ERR2 = SP2 - TC2
      ERR1 = SP1 - TC1
!-----Integral control action
      DIC2 = KI2*ERR2
      DIC1 = KI1*ERR1
!-----Proportional control action
      C2 = IC2 + KP2*ERR2
      C1 = IC1 + KP1*ERR1
!-----Calculate cooling HTC
      HTC2 = 25.4*C2/((TC2-AMBT)*(SDZ(CFLAG2)-SDZ(CFLAG2-1)))
      HTC1 = 25.4*C1/((TC1-AMBT)*(SDZ(CFLAG1)-SDZ(CFLAG1-1)))
      DO 23 I=1,NEL
      IF(I.LT.MPFLAG) CONT(I) = 11.81*HTC1*DELZ(I)*(TS(I)-AMBT)
      IF(I.EQ.MPFLAG) CONT(I) = 11.81*HTC2*(SDELZ(1+MPFLAG)-MPB)* &
               (TS(I)-AMBT)
      IF(I.GT.MPFLAG) CONT(I) = 11.81*HTC2*DELZ(I)*(TS(I)-AMBT)
      IF(NCONT.EQ.0) CONT(I)=0.0
   23..CONTINUE


!-----Total cooling duty initiation
      SCDUTY = 0.0
      DO 22 I=1,NEL
!-----Mass balance of A
!-----Diffusion term
      IF(I.EQ.1.OR.I.EQ.NEL) D2ADZ2 = 0.0
      IF(I.EQ.2) D2ADZ2=((XAIN-XA(I))/(2.0*DZ(I))-(XA(I-1) &
       -XA(I+1))/(2.0*DZ(I+1)))/DELZ(I)
      IF(I.GT.2.AND.I.LT.NEL) D2ADZ2 = ((XA(I-2)-XA(I))/(2.0&
       *DZ(I)) - (XA(I-1)-XA(I+1))/(2.0*DZ(I+1)))/DELZ(I)
!-----Velocity term
      DADZ =. (XA(I-1) - XA(I))/DELZ(I)
      IF(I.EQ.1) DADZ = (XAIN - XA(1))/DELZ(I)
```

```
          DXA(I) = 4.57E-4*D2ADZ2 + 0.04035*DADZ - ENDRR(I)
          IF(XA(I).LE.0.0.AND.DXA(I).LT.0.0) DXA(I) = 0.0
          IF(XA(I).LT.0.0) XA(I) = 0.0
          IF(XA(I).EQ.0.0) XA(I+1)=0.0


!-----Mass balance of B
!     XB is strictly a state variable which may be defined in a manner
!     similar to that for XA and XC.  For simplicity, XB is defined
!     from the fact that XA+XB+XC=1.0.
!-----Algebraic form of B mass balance equations
!-----   to preserve stoichiometry.
          XB(I) = 1.00000 - XA(I) - XC(I)


!-----Mass balance of C
!-----Diffusion term
          IF(I.EQ.1.OR.I.EQ.NEL) D2CDZ2 = 0.0
          IF(I.EQ.2) D2CDZ2 = ((XCIN-XC(I))/(2.0*DZ(I)) - &
          (XC(I-1)-XC(I+1))/(2.0*DZ(I+1)))/DELZ(I)
          IF(I.GT.2.AND.I.LT.NEL) D2CDZ2 = ((XC(I-2)-XC(I))/(2.0* &
          DZ(I)) - (XC(I-1)-XC(I+1))/(2.0*DZ(I+1)))/DELZ(I)
!-----Velocity term
          DCDZ = (XC(I-1)-XC(I))/DELZ(I)
          IF(I.EQ.1) DCDZ = (XCIN-XC(1))/DELZ(I)

          DXC(I) = 4.57E-4*D2CDZ2 + 0.04035*DCDZ + EXRR(I)
          IF(XC(I).GE.1.0.AND.DXC(I).GT.0.0) DXC(I) = 0.0
          IF(XC(I).GT.1.0) XC(I) = 1.0
!-----Solids heat balance
!-----Diffusion term
          IF(I.EQ.1.OR.I.EQ.NEL) D2SDZ2 = 0.0
          IF(I.EQ.2) D2SDZ2=((TSIN-TS(2))/(2.0*DZ(I))-(TS(1)-TS(3))/&
          (2.0*DZ(I+1)))/DELZ(I)
          IF(I.GT.2.AND.I.LT.NEL) D2SDZ2=((TS(I-2)-TS(I))/(2.0*DZ(I))-&
          (TS(I-1)-TS(I+1))/(2.0*DZ(I+1)))/DELZ(I)
!-----Velocity term
          IF(I.GT.1) DTSDZ = (TS(I) - TS(I-1))/DELZ(I)
          IF(I.EQ.1) DTSDZ = (TS(I) - TSIN)/DELZ(I)
!-----Average solids temperature
          IF(I.GT.1) TSMEAN = 0.5*(TS(I) + TS(I-1))
          IF(I.EQ.1) TSMEAN = 0.5*(TS(I) + TSIN)
!-----Average gas temperature
          TGMEAN = 0.5*(TG(I)+TG(I+1))
          IF(I.EQ.NEL) TGMEAN = 0.5*(TG(I) + TGIN)
!-----Fourth power wall temperatures
          TW4 = TW(I)*TW(I)*TW(I)*TW(I)
!-----Fourth power solids temperatures
          TP4 = TSMEAN*TSMEAN*TSMEAN*TSMEAN
!-----Heat balance
          DTS(I) = 4.57E-4*D2SDZ2 - 0.04035*DTSDZ + 1.235*EXRR(I) &
          + 0.2895*(TW(I) - TSMEAN) + 3.9*(TGMEAN-TSMEAN) + 0.1995* &
          (TW4-TP4) - 0.3304*ENDRR(I)
          IF(TS(I).LT.0.0) DTS(I)= 0.0
          IF(TS(I).LT.0.0) TS(I) = 1.0E-9
```

```
!-----WALL HEAT BALANCE
!-----Total cooling duty
      SCDUTY = SCDUTY + CONT(I)
!-----Diffusion term
      IF(I.EQ.1.OR.I.EQ.NEL) D2WDZ2 = 0.0
      IF(I.GT.1.AND.I.LT.NEL) D2WDZ2 = ((TW(I-1)-TW(I))/DZ(I) &
      - (TW(I)-TW(I+1))/DZ(I+1))/DELZ(2)
!-----Heat balance
      DTW(I) = 2.93E-5*D2WDZ2 + 0.0909*(TSMEAN-TW(I)) &
      + 0.0626*(TP4-TW4) + 0.1026*(TGMEAN-TW(I)) + CONT(I)
   22..CONTINUE
   END  !OF PROCEDURAL
      TS = INTVC(DTS,TS0)
      TW = INTVC(DTW,TW0)
      XA = INTVC(DXA,XA0)
      XC = INTVC(DXC,XC0)
      IC1 = INTEG(DIC1,IC10)
      IC2 = INTEG(DIC2,IC20)
   END  !OF DERIVATIVE


   DISCRETE SPPROF
!----------OUTPUT SPATIAL PROFILE AT INTERVALS GIVEN BY SPINT
! Data is written in basic tabular format. It may need some editing
! to render it suitable for input to a spreadsheet or other graph
! plotting package for production of spatial profile plots.
      IF(NEVENT.EQ.1) WRITE(11,1000)(SDELZ(I),I=1,NEL+1)
   1000..FORMAT(21E11.4)
      WRITE(11,1001)T,TSIN,(TS(J),J=1,NEL)
   1001..FORMAT(22E11.4)
      WRITE(11,1001)T,(TG(J),J=1,NEL),TGIN
      WRITE(11,1001)T,XAIN,(XA(J),J=1,NEL)
      WRITE(11,1001)T,XBIN,(XB(J),J=1,NEL)
      WRITE(11,1001)T,XCIN,(XC(J),J=1,NEL)
      NEVENT=NEVENT+1
      SCHEDULE SPPROF .AT. T+SPINT
   END  !OF DISCRETE


   TERMT(T.GE.TSTOP)
   CINT = 1.0
   ALGORITHM IALG=9
END  !OF DYNAMIC


TERMINAL
   CLOSE (UNIT=11)
END  !OF TERMINAL


END  !OF PROGRAM
```

A.1.16   REACTOR.CMD

Copyright © B S Bennett, 1994

```
SET HVDPRN = .T.
SET CALPLT = .T.
SET STRPLT = .F.
SET SYMCPL = .T.
SET NPCCPL = 5
SET TITLE = 'COUNTER-CURRENT TUBULAR REACTOR MODEL'
OUTPUT  T, TS(2),TS(4),TS(6),TS(8),TS(10),TS(12),TS(14),TS(16), &
    TS(18),TW(2),TW(4),TW(6),TW(8),TW(10),TW(12),TW(14),TW(16),TW(18)
PREPAR T,TS(1),TG(1),TS(20),TG(20),TC1,TC2,SP1,SP2

PROCED GO
   SPARE
   START
   SPARE
END
! The following procedures are supplied to provide some time domain
!   plots after the run has been completed
PROCED PLOT1
   PLOT /XAXIS=T, /XTAG='(MINUTES)', /XLO=0.0, /XHI=100.0, /COLOR=0, &
        TS(1),/TAG='(Solids Temp.Sect.5)',/CHAR=1, /STYLE=0, &
        TG(1),/TAG='(Gas Temp.Sect.1)', /STYLE=1, &
        TS(20),/TAG='(Solids Temp.Sect.20)', /STYLE=2, &
        TG(20),/TAG='(Gas Temp.Sect.20)', /STYLE=3
END
PROCED PLOT2
   PLOT /XAXIS=T, /XTAG='(MINUTES)', /XLO=0.0, /XHI=100.0, &
      /COLOR=0, /LO=0.5,/HI=1.5, &
      SP1,/TAG='Temperature 1 Set Point',/CHAR=1, /STYLE=1, &
      TC1,/TAG='Controlled Shell Temperature 1', /STYLE=0, &
      SP2,/TAG='Temperature 2 Set Point', /STYLE=1, &
      TC2,/TAG='Controlled Shell Temperature 2', /STYLE=0, /SAME
END
PROCED PLOT3
   PLOT /XAXIS=T, /XTAG='(MINUTES)', /XLO=0.0, /XHI=20.0, &
      /COLOR=0, /LO=0.5, /HI=1.5, &
      SP2,/TAG='Temperature 2 Set Point', /CHAR=1, /STYLE=1, &
      TC2,/TAG='Controlled Shell Temperature 2', /STYLE=0, /SAME
END
PROCED PLOT4
   PLOT /XAXIS=T, /XTAG='(MINUTES)', /XLO=0.0, /XHI=20.0, &
      /COLOR=0, /LO=0.5, /HI=0.55, &
      SP1,/TAG='Temperature 1 Set Point', /CHAR=1, /STYLE=1, &
      TC1,/TAG='Controlled Shell Temperature 1', /STYLE=0, /SAME
END
SET CMD = 0
```

A.1.17 SAMPLE.CSL

Copyright © B S Bennett, 1994

```
PROGRAM SIMPLE SAMPLED-DATA FEEDBACK SYSTEM

INITIAL
    CONSTANT X10=0.0, X20=0.0, SETPT=1.0, TP=0.8, GAIN1=10.0, &
             GAIN2=1.0, POLE=3.5, SERROR=0.0
!----------Initialise the discrete section that handles sampling
    SCHEDULE SAMPLE .AT. 0.0
END  !OF INITIAL

DYNAMIC
   DERIVATIVE
!----------Compute continuous error
   CERROR = SETPT - X2
!----------Integrator output
   X1 = INTEG(GAIN1*SERROR,X10)
!----------Final output
   X2 = INTEG(X1 - POLE*X2,X20)
   END  !OF DERIVATIVE

   DISCRETE SAMPLE
!----------Log all output data before processing event
   CALL LOGD(.FALSE.)
!----------Sample the continuous error
   SERROR = CERROR
!----------Log all output data after processing event
   CALL LOGD(.FALSE.)
!----------Schedule next time at which event SAMPLE will be processed
   SCHEDULE SAMPLE .AT. T+TP
   END  !OF DISCRETE

   TSTOP = 10.0
   CINT = 0.1
   ALGORITHM IALG=9
   TERMT(T.GE.TSTOP)
END  !OF DYNAMIC

END  !OF PROGRAM
```

A.1.18 SAMPLE.CMD

Copyright © B S Bennett, 1994

```
SET STRPLT=.F.
SET CALPLT=.T.
SET SYMCPL=.T.
SET NPCCPL=18
SET TITLE = 'SIMPLE SAMPLED-DATA SYSTEM'
OUTPUT /LO=-1,/HI=8,T,/LO=0,/HI=10,X1,/CHAR=1,X2,CERROR,SERROR
PREPAR T,X1,X2,CERROR,SERROR
```

```
PROCED GO
SPARE
START/PLOT
SPARE
END
PROCED PLOT1
PLOT /XAXIS=T,X1,/COLOR=0,/CHAR=1,X2,CERROR
END
PROCED PLOT2
PLOT CERROR,/COLOR=0,/CHAR=1,SERROR
END
PROCED PLOT3
PLOT X1,/COLOR=0,/CHAR=1,X2,CERROR,SERROR
END
PROCED PLOT4
PLOT X1,/COLOR=0,/CHAR=1,X2
END
PROCED PLOT5
PLOT /LO=-1.0,/HI=8.0,X1,/COLOR=0,/CHAR=1,X2,CERROR,SERROR,/SAME
END
PROCED PLOT6
PLOT /XAXIS=X1,/XTAG='(X1)',/XLO=-1.0,/XHI=8.0,/COLOR=0,X2,/CHAR=1
END
PROCED SETSTR
SET CALPLT=.F.
SET STRPLT=.T.
SET SYMSPL=.T.
SET NPCSPL=18
END
PROCED SETCAL
SET STRPLT=.F.
SET CALPLT=.T.
SET SYMCPL=.T.
SET NPCCPL=18
END
SET CMD=0
```

A.1.19 SAWTL.CSL

Copyright © B S Bennett, 1994

```
PROGRAM  SAWTOOTH GENERATOR - LATCH OPERATION

CONSTANT STMAX=1.0, STMIN=0.0, RAMPUP=3.0, RAMPDN=-1.5
INITIAL
   XIC=0.0
   LATCH=1  !A latch set to initiate positive slope
   SLOPE=RAMPUP  !Initial slope
END  ! OF INITIAL

DYNAMIC
```

```
DERIVATIVE
!---------The following procedural section tests for hitting the
!---------limits of the sawtooth waveform, and allows change of
!---------slope if this does happen.
    PROCEDURAL(SLOPE = X,LATCH,RAMPUP,RAMPDN,STMAX,STMIN)
        IF(X.LE.STMIN.AND.LATCH.EQ.-1) LATCH=1
        IF(X.GE.STMAX.AND.LATCH.EQ.1) LATCH=-1
        IF(LATCH.EQ.1) SLOPE=RAMPUP
        IF(LATCH.EQ.-1) SLOPE=RAMPDN
    END   ! OF PROCEDURAL
    XD = SLOPE
    X = INTEG(XD,XIC)
END   ! OF DERIVATIVE

CINTERVAL CINT = 0.1
CONSTANT TSTOP=10.0
TERMT(T.GE.TSTOP)
ALGORITHM IALG=9
END ! OF DYNAMIC

END   ! OF PROGRAM
```

A.1.20 SAWTL.CMD

Copyright © B S Bennett, 1994

```
SET HVDPRN=.T.
SET CALPLT=.F.
SET STRPLT=.T.
SET SYMSPL=.F.
SET NPCSPL=20
SET TITLE='SAWTOOTH WAVE FORM - LATCH OPERATION'
OUTPUT /LO=-2,/HI=4,T,/LO=0,/HI=10,X,XD
PREPAR T,X,XD
PROCED GO
    SPARE
    START/PLOT
    SPARE
END
PROCED PLOT1
PLOT /XTAG='(SEC)', X, /TAG='(SAWTOOTH)',/COLOR=0,/LO=0,/HI=1, &
     XD, /TAG='(SLOPE)',/LO=-2,/HI=4
END
SET CMD=0
```

A.1.21   SAWTEV.CSL

Copyright © B S Bennett, 1994

```
PROGRAM  SAWTOOTH GENERATOR - WITH EVENT PROCESSING

    CONSTANT STMAX=1.0,STMIN=0.0,RAMPUP=3.0,RAMPDN=-1.5

INITIAL
    XIC = 0.0
    SCHEDULE UP .AT. 0.0  !Put first event in queue to start waveform
END  ! OF INITIAL

DYNAMIC

    DERIVATIVE
        X = INTEG(XD,XIC)
    END  ! OF DERIVATIVE
!--------Event to set up positive ramp and to schedule the next event
    DISCRETE UP
        CALL LOGD(.FALSE.)  !To cause output before event processing
        SLOPE=RAMPUP
        EVNTUP=(STMAX-STMIN)/SLOPE
        XD = SLOPE
        CALL LOGD(.FALSE.)  !To cause output after event processing
        SCHEDULE DOWN .AT. T+EVNTUP  !Schedule next event (DOWN)
    END  !OF DISCRETE UP
!--------Event to set up negative ramp and to schedule event above
    DISCRETE DOWN
        CALL LOGD(.FALSE.)  !To cause output before event processing
        SLOPE=RAMPDN
        EVNTDN=(STMIN-STMAX)/SLOPE
        XD = SLOPE
        CALL LOGD(.FALSE.)  !To cause output after event processing
        SCHEDULE UP .AT. T+EVNTDN  !Schedule next event (UP)
    END  !OF DISCRETE DOWN

    CINTERVAL CINT = 0.1
    CONSTANT TSTOP=10.0
    TERMT(T.GE.TSTOP)
    ALGORITHM IALG=9
END  ! OF DYNAMIC

END  ! OF PROGRAM
```

A.1.22   SAWTEV.CMD

Copyright © B S Bennett, 1994

```
SET HVDPRN=.T.
SET CALPLT=.F.
SET STRPLT=.T.
SET SYMSPL=.F.
```

```
SET GLTPLT=210
SET ALCPLT=.F.
SET NPCSPL=20
SET TITLE='SAWTOOTH WAVE FORM - EVENT PROCESSING'
PREPAR T,X,XD
OUTPUT T,/LO=0,/HI=10,X,/LO=0,/HI=1,/TYPE=000, &
    XD,/LO=-2,/HI=4,/TYPE=100
PROCED GO
   SPARE
   START/PLOT
   SPARE
END
PROCED PLOT1
PLOT /XTAG='(SEC)', X, /TAG='(SAWTOOTH)',/TYPE=000,/LO=0,/HI=1, &
     XD, /TAG='(SLOPE)',/TYPE=000,/LO=-2,/HI=4
END
SET CMD=0
```

A.1.23 WHALE.CSL

Copyright © B S Bennett, 1994

```
PROGRAM    WHALE POPULATION DYNAMICS

INITIAL
   CONSTANT FACTOR=0.0, YOUNG0=15000, ADULT0=15000, ELDER0=10000

END  !OF INITIAL

DYNAMIC

   DERIVATIVE
   !---------Annual Rate of Change of Young Whale Population
   YOUNGD = 0.205*ADULT + 0.225*ELDER - 0.36*YOUNG - FACTOR*YOUNG
   !---------Population of Young Whales
   YOUNG = INTEG(YOUNGD,YOUNG0)
   !---------Annual Rate of Change of Adult Whale Population
   ADULTD = 0.25*YOUNG - 0.235*ADULT - FACTOR*ADULT
   !---------Population of Adult Whales
   ADULT = INTEG(ADULTD,ADULT0)
   !---------Annual Rate of Change of Elderly Whale Population
   ELDERD = 0.125*ADULT - 0.18*ELDER - FACTOR*ELDER
   !---------Population of Elderly Whales
   ELDER = INTEG(ELDERD,ELDER0)

   END  !OF DERIVATIVE

   TSTOP = 50
   CINT=0.5
   ALGORITHM IALG=9
   TERMT(T.GE.TSTOP)

END  !OF DYNAMIC

END  !OF PROGRAM
```

A.1.24   WHALE.CMD

Copyright © B S Bennett, 1994

```
SET STRPLT=.F.
SET CALPLT=.T.
SET SYMCPL=.T.
SET NPCCPL=18
SET TITLE = 'WHALE POPULATION DYNAMICS'
OUTPUT /LO=0,/HI=25000,T,/HI=50,YOUNG,/CHAR=1,ADULT,ELDER
PREPAR T,YOUNG,ADULT,ELDER
PROCED GO
SPARE
START/PLOT
SPARE
END
PROCED PLOT1
PLOT /XAXIS=T,/XTAG='(YEARS)',YOUNG,/COLOR=0,/CHAR=1,/HI=25000, &
     /LO=0,ADULT,ELDER,/SAME
END
PROCED PLOT2
PLOT YOUNG,/COLOR=0,/CHAR=1,ADULT,ELDER,
END
SET CMD=0
```

Appendix A.2

MATSIM Models Used in this Book

The files listed in this appendix are M-files that run under MATLAB 3.5. The files go together in pairs, each pair forming a simulation model that is presented as an example in the book. In connection with the execution of these models, the following notes may be useful:

- One file of each pair is devoted solely to the model equations coded in the MATLAB language. This includes any event processing that is part of the model operation. The names of these M-files end in the letter B, for example BEAMB.M, MULTRB.M.
- The second file of each pair is an initialization file that sets up all model constants, parameters and initial conditions, and contains MATLAB statements for post-run plotting. Run-time plotting is handled by the MATSIM routines, provided that it is enabled in the initialization file. These files have names ending in IN, for example BEAMIN.M, MULTRIN.M.

A.2.1 BEAMIN.M

Copyright © B S Bennett, 1994

```
1.  % This is a simulation of the well-known BENDING BEAM PROBLEM
2.  % to test iterative operation.  The standard fourth-order
3.  % equation for beam deflection, EI(d4y/dx4) = -w is decomposed
4.  % into 4 first-order ODE's, where E=5.0E9, I=2.0E-2 and w=100.
5.  % With the beam straddling the space between two walls, and
6.  % embedded at both ends, this becomes a two-point boundary-
7.  % value problem, where the boundary conditions on the beam
8.  % slope and deflection dictate the initial conditions on the
9.  % shearing force and bending moment.

10. % Declare all global variables required in the simulation
11.     global h_min iter_ate t_out x_out y_out
12.     global run_plot xy_plot xy_size abs_xy a_a b_b x_label
13.     global points t_last y_last x_last v_plot
14.     global x_var y_var x_zero e_vent per_turb
```

```
15. % Set run conditions
16.    t0 = 0;                              % simulation start time
17.    run_plot = 1;                        % enable run-time plotting
18.    tfin = 100;                          % simulation finish time
19.    x0 = [0.0,0.0,0.0,0.0];              % initial conditions
20.    x_zero = x0;                         % save initial values
21.    per_turb = 1;                        % iteration counter
22.    cint = (tfin - t0)/25;               % output interval
23.    h_min = cint/1000;                   % minimum allowable step size
24.    v_plot = [t0,tfin,-1e-4,1e-4;t0,tfin,-0.001,0.0005;t0,tfin,-0.01,0.01;
25.             t0,tfin,-0.3,0.1];          % run-time plot scaling
26.    xy_plot = ['x4';'x3';'x2';'x1'];     % run-time plot variable names
27.    x_label = 'length';                  % independent variable name
28.    [dum1] = decode(t0);                 % run-time plot parameters

29. % Start the simulation
30.    [t,x] = feval('dynamb','ode23b','beamb',t0,tfin,x0,cint);

31. % Return graphics screen to default state and produce post-run plots
32.    shg,pause,hold off
33.    subplot
34.    plot(t_out,x_out(:,1),t_out,x_out(:,2),'--',t_out,x_out(:,3),'-.', ...
35.       t_out,x_out(:,4),':'),xlabel(x_label)
36.     title('bending beam simulation'),pause
37.    subplot (221)
38.    plot(t_out,x_out(:,4)),xlabel(x_label),title('shear force')
39.    subplot (222)
40.    plot(t_out,x_out(:,3)),xlabel(x_label),title('bending moment')
41.    subplot(223)
42.    plot(t_out,x_out(:,2)),xlabel(x_label),title('beam slope')
43.    subplot(224)
44.    plot(t_out,x_out(:,1)),xlabel(x_label),title('beam deflection'),pause
45.    clg
```

A.2.2 BEAMB.M

Copyright © B S Bennett, 1994

```
1. % This is the model for the BENDING BEAM PROBLEM.  The largest part of
2. % this model is the terminal region where the initial conditions are
3. % adjusted for subsequent runs, and the sensitivity matrix computed to
4. % determine the initial values likely to yield the desired boundary
5. % values.  The integration is carried out along the length of the beam,
6. % for which time is substituted as the independent variable.  An 'event'
7. % is set up to occur along the beam (for the application of a point load),
8. % but in this listing, it is well outside the total beam length, and is
9. % therefore a dummy event.

10. function dx = beamb(t,x,control)

11.    if control == 1                      % INITIAL REGION
12.        e_vent(1) = 200;                 % set up a dummy event 'time'
13.        x_var = x_zero;                  % initialise working x-array
14.    end
```

```
15.    if control == 2                    % DYNAMIC REGION
16.        dx(4) = -100/5e9/2e-2;         % equation for shearing force
17.        dx(3) = x(4);                  % equation for bending moment
18.        dx(2) = x(3);                  % equation for beam slope
19.        dx(1) = x(2);                  % equation for beam deflection
20.    end

21.    if control == 5                    % PROCESS EVENT OF CLASS 1
22.        x_var(4) = x_var(4) - 10000/5e9/2e-2;  % apply point load
23.        e_vent(1) = 0;                 % reset event to zero
24.    end

25.    if control == 3                    % TERMINAL REGION
26.        if abs(x(2))>5e-4|abs(x(1))>5e-3  % if final values of slope and
27.                                        % deflection are not zero, THEN
28.            if per_turb == 3           % for 3rd pass:
29.                y_var(5:6) = x_var(1:2);   % save these final values
30.                a = [0,0;0,0];             % initialise (2 x 2)
31.                a(1,1) = (y_var(5)-y_var(1))/5e-5;  % sensitivity matrix
32.                a(1,2) = (y_var(3)-y_var(1))/5e-6;  % and compute Jacobian
33.                a(2,1) = (y_var(6)-y_var(2))/5e-5;  % elements from
34.                a(2,2) = (y_var(4)-y_var(2))/5e-6;  % perturbation figures
35.                b = inv(a);                % invert sensitivity matrix
36.                x_zero(3) = y_var(7) - b(1,1)*y_var(1) - b(1,2)*y_var(2);
37.                x_zero(4) = y_var(8) - b(2,1)*y_var(1) - b(2,2)*y_var(2);
38.            end                        % compute desired initial values
39.            if per_turb == 2;          % for 2nd pass:
40.                y_var(3:4) = x_var(1:2);   % save these final values
41.                x_zero(3) = y_var(7) + 5e-5;  % reset initial shear force
42.                x_zero(4) = y_var(8);      % to original value (zero)
43.                per_turb = 3;              % perturb initial bending
44.            end                        % moment for next run
45.            if per_turb == 1           % for 1st pass:
46.                y_var(1:2) = x_var(1:2);   % save these final values
47.                y_var(7:8) = x_zero(3:4);  % save the corresponding
48.                x_zero(4) = y_var(8) + 5e-6;  % initial values, and perturb
49.                per_turb = 2;              % initial shear force for
50.            end                        % next run
51.            iter_ate = 1;              % enable an iteration to take place
52.        else                           %    OTHERWISE
53.            iter_ate = 0;              % disable iteration (because final
54.        end                            % values are close enough to zero)
55.    end
```

A.2.3 CONTIN.M

```
1. % This is a simulation of a SIMPLE CONTINUOUS FEEDBACK SYSTEM

2. % Declare all global variables required in the simulation
3.    global h_min iter_ate t_out x_out y_out
```

```
4.     global run_plot xy_plot xy_size abs_xy a_a b_b x_label
5.     global points t_last y_last x_last v_plot
6.     global x_var y_var x_zero e_vent

7. %Set run conditions
8.     t0 = 0;                             % simulation start time
9.     run_plot = 1;                       % enable run-time plotting
10.    tfin = 10.0;                        % simulation finish time
11.    x0 = [0.0,0.0];                     % initial conditions
12.    x_zero = x0;                        % save initial values
13.    cint = (tfin - t0)/100;             % output interval
14.    h_min = cint/1000;                  % minimum allowable step size
15.    v_plot = [t0,tfin,-4,4;t0,tfin,0,2;t0,tfin,-1,1;t0,tfin,0,2];
16.                                        % run-time plot scaling
17.    xy_plot = ['x1';'x2';'y1';'y2'];    % run-time plot variable names
18.    x_label = 'time';                   % independent variable name
19.    [dum1] = decode(t0);             % run-time plot parameters

20. % Start the simulation
21.    [t,x] = feval('dynamb','ode23b','contb',t0,tfin,x0,cint);

22. % Return graphics screen to default state and produce post-run plots
23.    shg,pause,hold off
24.    subplot
25.    plot(t_out,x_out(:,1),t_out,x_out(:,2),':',t_out,y_out(:,1),'--')
26.      title('Simple F/B Control System'),pause
27.    subplot (211)
28.    plot(t_out,x_out(:,1),t_out,x_out(:,2),':')
29.      title('State variables x1 (intgl err) and x2 (output)'),pause
30.    subplot (212)
31.    plot(t_out,y_out(:,1),t_out,y_out(:,2),':')
32.      title('y1 (error) and y2 (output)'),pause
33.    subplot
34.    plot(x_out(:,1),x_out(:,2),x_out(:,1),y_out(:,1),x_out(:,1),y_out(:,2),':')
35.      title('phase-plane plots'),pause
36.    clg
```

A.2.4 CONTB.M

```
1. % This is the model for a SIMPLE CONTINUOUS FEEDBACK SYSTEM.
2. % There are two state (x-)variables and two algebraic (y-)variables

3. function dx = contb(t,x,control)

4.        y_var(1) = 1.0 - x(2);          % continuous error
5.        dx(1) = 10.0*y_var(1);          % integral of error
6.        dx(2) = x(1) - 0.5*x(2);        % first-order lag
7.        y_var(2) = 1.0-y_var(1);        % algebraic output equation
```

A.2.5 HLIMIT.M

Copyright © B S Bennett, 1994

```
1.  % This is a simple hard-limit control function implemented
2.  % for use in, for example, rate-limiting a control valve movement

3.  function yout = hlimit(yin,llimit,ulimit)

4.      if ulimit>llimit      % if data entered in wrong order
5.          ul = ulimit;      % swap around, otherwise implement
6.          ll = llimit;      % as given
7.      else
8.          ul = llimit;
9.          ll = ulimit;
10.     end

11.     yout = yin;           % set output equal to input, but if
12.     if yin>ul             % greater than the upper limit, set
13.         yout = ul;        % equal to the upper limit
14.     end                   %   or
15.     if yin<ll             % if less than the lower limit, set
16.         yout = ll;        % equal to the lower limit.
17.     end
```

A.2.6 MULTRIN.M

Copyright © B S Bennett, 1994

```
1.  % This is a simulation of a MULTI-RATE SAMPLED-DATA CONTROL SYSTEM
2.  % to test multi-rate sampling action as mechanised by
3.  % event-processing.

4.  % Declare all global variables required in the simulation
5.      global h_min iter_ate t_out x_out y_out
6.      global run_plot xy_plot xy_size abs_xy a_a b_b x_label
7.      global points t_last y_last x_last v_plot
8.      global x_var y_var x_zero e_vent t_samp co_eff t_con

9.  % Set run conditions
10.     t0 = 0;                         % simulation start time
11.     run_plot = 1;                   % enable run-time plotting
12.     tfin = 5;                       % simulation finish time
13.     x0 = [0,0,0];                   % initial conditions
14.     x_zero = x0;                    % save initial values
15.     y_var = zeros(1,10);           % set up initialised y-array
16.     t_samp = [0.2,0.6];            % two sampling periods
17.     t_con = [0.5,0.5,0.1,0.1];     % double phase-advance time constants
18.     cint = (tfin - t0)/25;         % output interval
19.     h_min = cint/1000;             % minimum allowable step size
20.     v_plot = [t0,tfin,0,10;t0,tfin,0,1.5;t0,tfin,-0.5,1;t0,tfin,-0.5,1];
21.                                     % run-time plot scaling
```

```
22.    xy_plot = ['x1';'x3';'y1';'y8'];        % run-time plot variable names
23.    x_label = 'time';                       % independent variable name
24.    [dum1] = decode(t0);                    % run-time plot parameters


25. % Start the simulation
26.    [t,x] = feval('dynamb','ode23b','multrb',t0,tfin,x0,cint);


27. % Return graphics screen to default state and produce post-run plots
28.    shg,pause,hold off
29.    subplot
30.    plot(t_out,x_out(:,3),t_out,y_out(:,1),':',t_out,y_out(:,8),'--'),
31.        xlabel(x_label),title('Multi-rate S/D F/B Control System'),pause;
32.    subplot (211)
33.    plot(t_out,x_out(:,1),t_out,x_out(:,3),':'),xlabel(x_label)
34.        title('State variables x1 (valve pos.) and x3 (output)'),pause;
35.    subplot (212)
36.    plot(t_out,y_out(:,1),t_out,y_out(:,8),':'),xlabel(x_label)
37.        title('y1 (error) and y8 (sampled error)'),pause;
38.    subplot
39.    plot(t_out,x_out(:,1),t_out,y_out(:,3),':',t_out,y_out(:,4),'--'),
40.        xlabel(x_label)
41.        title('x1 (valve pos.), y3 (rate demand), y4 (rate limit)'),pause;
42.    plot(t_out,y_out(:,2),t_out,y_out(:,5),':'),xlabel(x_label)
43.        title('y2 (sampled compensator O/P) and y5 (compensator O/P)'),pause;
44.    clg
```

A.2.7 MULTRB.M

```
1. % This is a model for a MULTI-RATE SAMPLED-DATA CONTROL SYSTEM.  There are
2. % two sample rates, and the control is by a discretised double phase
3. % advance compensator.

4. function dx = multrb(t,x,control)

5.    if control == 1                          % INITIAL REGION
6.                                             % compute discrete control parameters
7.        ts2 = 1.0/(t_samp(1)*t_samp(1));
8.        a1 = (t_samp(1)+t_con(3))*(t_samp(1)+t_con(4))*ts2;
9.        b1 = ((t_con(3)+t_con(4))*t_samp(1)+2*t_con(3)*t_con(4))*ts2;
10.        c1 = t_con(3)*t_con(4)*ts2;
11.        a2 = (t_samp(1)+t_con(1))*(t_samp(1)+t_con(2))*ts2;
12.        b2 = ((t_con(1)+t_con(2))*t_samp(1)+2*t_con(1)*t_con(2))*ts2;
13.        c2 = t_con(1)*t_con(2)*ts2;
14.        co_eff(1) = 2*a2/a1;
15.        co_eff(2) = b2/a2;
16.        co_eff(3) = c2/a2;
17.        co_eff(4) = b1/a1;
18.        co_eff(5) = c1/a1;
19.    end
```

```
20.    if control == 2                    % DYNAMIC REGION
21.        y_var(1) = 1.0 - x(3);             % continuous error
22.        y_var(3) = 2.0*y_var(2);           % control valve rate demand
23.        y_var(4) = hlimit(y_var(3),-15,15); % control valve rate limit
24.        dx(1) = y_var(4);                  % control valve position
25.        dx(2) = x(1) - 4*x(2) - 8*x(3);    % second order transfer function
26.        dx(3) = x(2);                      % response to valve position
27.    end

28.    if control == 5                    % PROCESS EVENT OF CLASS 1
29.        y_var(7) = y_var(6);               % Move previous samples down
30.        y_var(6) = y_var(5);               % shift register, and sample
31.        y_var(10) = y_var(9);              % the continuous error.
32.        y_var(9) = y_var(8);               % Compute the output of the
33.        y_var(8) = y_var(1);               % control algorithm
34.        y_var(5) = co_eff(1)*(y_var(8) - co_eff(2)*y_var(9)   ...
35.            + co_eff(3)*y_var(10)) + co_eff(4)*y_var(6) - co_eff(5)*y_var(7);
36.        e_vent(1) = t + t_samp(1);         % set next event time in class
37.    end

38.    if control == 6                    % PROCESS EVENT OF CLASS 2
39.        y_var(2) = y_var(5);               % sample control algorithm output
40.        e_vent(2) = t + t_samp(2);         % set next event time in class
41.    end
```

A.2.8 MULTNIN.M

Copyright © B S Bennett, 1994

```
1. % This is a simulation of the same MULTI-RATE SAMPLED-DATA CONTROL SYSTEM
2. % as before, but WITH NOISE ON MEASURED OUTPUT.   This is an example of
3. % the use of event-processing to generate a form of band-limited noise
4. % by relatively simple means.   Run-time plotting is only possible with
5. % 386-MATLAB or MATLAB 4.0 because of memory constraints.
6. % Use is made of MATLAB functions to produce a power spectral
7. % density plot to demonstrate the quality of the noise function.
8. % Declare all global variables required in the simulation
9.    global h_min iter_ate t_out x_out y_out
10.   global run_plot xy_plot xy_size abs_xy a_a b_b x_label
11.   global points t_last y_last x_last v_plot
12.   global x_var y_var x_zero e_vent t_samp co_eff t_con t_yy y_yy

13. %Set run conditions
14.   t0 = 0;                            % simulation start time
15.   run_plot = 1;                      % enable run-time plotting
16.   tfin = 5;                          % simulation finish time
17.   x0 = [0,0,0];                      % initial conditions
18.   x_zero = x0;                       % save initial values
19.   y_var = zeros(1,11);               % set up initialised y-array
20.   t_samp = [0.2,0.6,0.05];           % three sampling periods
21.   t_con = [0.5,0.5,0.1,0.1];         % double phase-advance time constants
22.   cint = (tfin - t0)/25;             % output interval
23.   h_min = cint/1000;                 % minimum allowable step size
24.   v_plot = [t0,tfin,0,1.5;t0,tfin,-0.2,0.2;t0,tfin,-0.5,1.5;t0,tfin,-0.5,1.5];
```

```
25.                                      % run-time plot scaling
26.    xy_plot = ['x3 ';'y11';'y1 ';'y8 '];  % run-time plot variable names
27.    x_label = 'time';                 % independent variable name
28.    [dum1] = decode(t0);              % run-time plot parameters


29. % Start the simulation
30.    [t,x] = feval('dynamb','ode23b','multnb',t0,tfin,x0,cint);


31. % Return graphics screen to default state and produce post-run plots
32.    shg,pause,hold off
33.    subplot
34.    plot(t_out,x_out(:,3),t_out,y_out(:,1),':',t_out,y_out(:,8),'--'),
35.        xlabel(x_label)
36.        title('Multi-rate S/D F/B Control System with noise'),pause;
37.    subplot (211)
38.    plot(t_out,x_out(:,1),t_out,x_out(:,3),':'),xlabel(x_label)
39.        title('State variables x1 (valve pos.) and x3 (output)'),pause;
40.    subplot (212)
41.    plot(t_out,y_out(:,1),t_out,y_out(:,8),':'),xlabel(x_label)
42.        title('y1 (error with noise) and y8 (sampled error)'),pause;
43.    subplot
44.    plot(t_out,x_out(:,1),t_out,y_out(:,3),':',t_out,y_out(:,4),'--'),
45.        xlabel(x_label)
46.        title('x1 (valve pos.), y3 (rate demand), y4 (rate limit)'),pause;
47.    plot(t_out,y_out(:,2),t_out,y_out(:,5),':'),xlabel(x_label)
48.        title('y2 (sampled compensator O/P) and y5 (compensator O/P)'),pause;
49.    plot(t_out,y_out(:,11),t_out,y_out(:,1),t_out,y_out(:,8),':')
50.        xlabel(x_label)
51.        title('y11 (noise), y1 (error with noise) and y8 (sampled error)'),pause;


52. % Calculate and plot power spectral density for noise function
53.    sizeyy = fix(size(y_yy)/2)*2;
54.    YY = fft(y_yy,sizeyy(1));
55.    PYY = YY.*conj(YY)/sizeyy(1);
56.    f = 1000*(0:sizeyy(1)/2-1)/sizeyy(1);
57.    plot(f,PYY(1:sizeyy(1)/2)),xlabel('frequency'),ylabel('Pyy')
58.        title('power spectrum of noise'),pause;
59.    clg
```

A.2.9   MULTNB.M

Copyright © B S Bennett, 1994

```
1. % This is the same model for a MULTI-RATE SAMPLED-DATA CONTROL SYSTEM
2. % but with noise added to the output signal.   An extra event-processing
3. % section is set up to generate this noise.


4. function dx = multnb(t,x,control)


5.    if control == 1                     % INITIAL REGION
6.                                        % compute discrete control parameters
7.        ts2 = 1.0/(t_samp(1)*t_samp(1));
```

```
8.          a1 = (t_samp(1)+t_con(3))*(t_samp(1)+t_con(4))*ts2;
9.          b1 = ((t_con(3)+t_con(4))*t_samp(1)+2*t_con(3)*t_con(4))*ts2;
10.         c1 = t_con(3)*t_con(4)*ts2;
11.         a2 = (t_samp(1)+t_con(1))*(t_samp(1)+t_con(2))*ts2;
12.         b2 = ((t_con(1)+t_con(2))*t_samp(1)+2*t_con(1)*t_con(2))*ts2;
13.         c2 = t_con(1)*t_con(2)*ts2;
14.         co_eff(1) = 2*a2/a1;
15.         co_eff(2) = b2/a2;
16.         co_eff(3) = c2/a2;
17.         co_eff(4) = b1/a1;
18.         co_eff(5) = c1/a1;
19.     end
20.     if control == 2                    % DYNAMIC REGION
21.         y_var(1) = 1.0 - x(3) + y_var(11);   % continuous error (with noise)
22.         y_var(3) = 2.0*y_var(2);            % control valve rate demand
23.         y_var(4) = hlimit(y_var(3),-15,15); % control valve rate limit
24.         dx(1) = y_var(4);                   % control valve position
25.         dx(2) = x(1) - 4*x(2) - 8*x(3);     % second-order transfer function
26.         dx(3) = x(2);                       % response to valve position
27.     end
28.     if control == 5                    % PROCESS EVENT OF CLASS 1
29.         y_var(7) = y_var(6);                % Move previous samples down
30.         y_var(6) = y_var(5);                % shift register, and sample
31.         y_var(10) = y_var(9);               % the continuous error.
32.         y_var(9) = y_var(8);                % Compute the output of the
33.         y_var(8) = y_var(1);                % control algorithm
34.         y_var(5) = co_eff(1)*(y_var(8) - co_eff(2)*y_var(9)  ...
35.             + co_eff(3)*y_var(10)) + co_eff(4)*y_var(6) - co_eff(5)*y_var(7);
36.         e_vent(1) = t + t_samp(1);          % set next event time in class
37.     end
38.     if control == 6                    % PROCESS EVENT OF CLASS 2
39.         y_var(2) = y_var(5);                % sample control algorithm output
40.         e_vent(2) = t + t_samp(2);          % set next event time in class
41.     end
42.     if control == 7                    % PROCESS EVENT OF CLASS 3
43.         t_yy = [t_yy;t];                    % update time and noise arrays
44.         y_yy = [y_yy;y_var(11)];            % generate amplitude sample from
45.         y_var(11) = (rand(1) - 0.5)*0.02*3.4641;  % random number generator
46.         t_yy = [t_yy;t];                    % update time and noise arrays
47.         y_yy = [y_yy;y_var(11)];
48.         e_vent(3) = t + t_samp(3);          % set next event time in class
49.     end
```

A.2.10 SAMPIN.M

```
1. % This is a simulation of a SIMPLE SAMPLED-DATA CONTROL SYSTEM
2. % to test single-rate sampling action as mechanised by
3. % event-processing.
```

```
4.  % Declare all global variables required in the simulation
5.      global h_min iter_ate t_out x_out y_out
6.      global run_plot xy_plot xy_size abs_xy a_a b_b x_label
7.      global points t_last y_last x_last v_plot
8.      global x_var y_var x_zero e_vent samp_period

9.  % Set run conditions
10.     t0 = 0;                             % simulation start time
11.     run_plot = 1;                       % enable run-time plotting
12.     tfin = 10;                          % simulation finish time
13.     x0 = [0.0,0.0];                     % initial conditions
14.     x_zero = x0;                        % save initial values
15.     samp_period = 0.8;                  % sampling period
16.     cint = (tfin - t0)/60;              % output interval
17.     h_min = cint/1000;                  % minimum allowable step size
18.     v_plot = [t0,tfin,0,8;t0,tfin,0,2;t0,tfin,-1,1;t0,tfin,0,2];
19.                                         % run-time plot scaling
20.     xy_plot = ['x1';'x2';'y1';'y3'];    % run-time plot variable names
21.     x_label = 'time';                   % independent variable name
22.     [dum1] = decode(t0);               % run-time plot parameters


23. % Start the simulation
24.     [t,x] = feval('dynamb','ode23b','sampb',t0,tfin,x0,cint);


25. % Return graphics screen to default state and produce post-run plots
26.     shg,pause,hold off
27.     subplot
28.     plot(t_out,x_out,t_out,y_out(:,1),'--',t_out,y_out(:,2),'-.',  ...
29.         t_out,y_out(:,3),':'),xlabel(x_label)
30.         title('Simple S/D F/B Control System'),pause
31.     subplot (211)
32.     plot(t_out,x_out(:,1),t_out,x_out(:,2),':'),xlabel(x_label)
33.         title('State variables x1 (intgl err) and x2 (output)'),pause
34.     subplot (212)
35.     plot(t_out,y_out(:,1),t_out,y_out(:,2),'--',t_out,y_out(:,3),':')
36.         xlabel(x_label)
37.         title('y1 (error), y2 (sampled error) and y3 (output)'),pause
38.     subplot
39.     plot(t_out,y_out(:,1),t_out,y_out(:,2),'--',t_out,y_out(:,3),':')
40.         xlabel(x_label)
41.         title('y1 (error), y2 (sampled error) and y3 (output)'),pause
42.     plot(x_out(:,1),x_out(:,2),x_out(:,1),y_out(:,3),':'),xlabel('x1')
43.         title('phase-plane plots'),pause
44.     clg
```

A.2.11 SAMPB.M

Copyright © B S Bennett, 1994

```
1.  % This is the same model as for the continuous control system implemented
2.  % as a SINGLE-RATE SAMPLED-DATA CONTROL SYSTEM

3.  function dx = sampb(t,x,control)
```

```
4.      if control == 1                     % INITIAL REGION
5.          y_var = [0,0];                  % initialise y-array
6.      end


7.      if control == 2                     % DYNAMIC REGION
8.          y_var(1) = 1.0 - x(2);          % continuous error
9.          dx(1) = 10.0*y_var(2);          % integral of sampled error
10.         dx(2) = x(1) - 3.5*x(2);        % first-order lag
11.         y_var(3) = 1.0 - y_var(1);      % algebraic output equation
12.     end


13.     if control == 5                     % PROCESS EVENT OF CLASS 1
14.         y_var(2) = y_var(1);            % sample continuous error
15.         e_vent(1) = t + samp_period;    % set next event time in class
16.     end
```

A.2.12 SAWTUNIN.M

Copyright © B S Bennett, 1994

```
1.  % This is a SAWTOOTH WAVEFORM GENERATOR, with simple logic applied
2.  % to bring about the change of direction  to show how step size is
3.  % reduced by discontinuities in such circumstances.

4.  % Declare all global variables required in the simulation
5.     global h_min iter_ate t_out x_out y_out
6.     global run_plot xy_plot xy_size abs_xy a_a b_b x_label
7.     global points t_last y_last x_last v_plot latch
8.     global x_var y_var e_vent

9.  % Set run conditions
10.    t0 = 0;                              % simulation start time
11.    run_plot = 1;                        % enable run-time plotting
12.    tfin = 6;                            % simulation finish time
13.    x0 = 0.0;                            % initial condition
14.    cint = (tfin - t0)/60;              % output interval
15.    h_min = cint/1000;                   % minimum allowable step size
16.    v_plot = [t0,tfin,0,1;t0,tfin,-2,4]; % run-time plot scaling
17.    xy_plot = ['x1';'y1'];               % run-time plot variable names
18.    x_label = 'time';                    % independent variable name
19.    [dum1] = decode(t0);                 % run-time plot parameters

20. % Start the simulation
21.    [t,x] = feval('dynamb','ode23b','sawtunb',t0,tfin,x0,cint);

22. % Return graphics screen to default state and produce post-run plots
23.    shg,pause,hold off
24.    subplot
25.    plot(t_out,x_out,t_out,y_out(:,1))
26.      title('sawtooth waveform - no event processing'),pause
27.    plot(t_out,x_out,'og')
28.      title('sawtooth waveform - no event processing'),pause
29.    plot(t_out,x_out,'.g')
30.      title('sawtooth waveform - no event processing'),pause
31.    clg
```

A.2.13   SAWTUNB.M

Copyright © B S Bennett, 1994

```
1.  % This is the model which is used to generate a SAWTOOTH WAVEFORM
2.  % without any event processing.  It relies solely on the use of simple
3.  % logic to cause the sawtooth derivative to change sign and value
4.  % once a peak or trough of the waveform was reached.  The subsequent
5.  % behaviour of the integration is typically that of successive step
6.  % size reduction to negotiate the discontinuities brought about by
7.  % slope reversal.

8.  function dx = sawtunb(t,x,control)

9.      if control == 1                    % INITIAL REGION
10.         latch = 1;                     % set latch for initial slope
11.     end

12.     if control == 2                    % DYNAMIC REGION
13.         if latch>0                     % if latch is set positive, then
14.             slope = 3.0;                   % set positive slope.  If
15.             if x(1) >= 1.0, slope = -1.5; end % sawtooth has exceeded peak
16.         else                               % value, change sign of slope
17.             slope = -1.5;                  % otherwise set negative slope. If
18.             if x(1) <= 0.0, slope = 3.0; end  % sawtooth has dropped below
19.         end                                % trough, set slope positive
20.         dx(1) = slope;                 % calculate waveform derivative
21.         y_var(1) = slope;              % save slope (for output)
22.     end

23.     if control == 4                    % END OF INTEGRATION STEP
24.         if x(1) > 0.9999, latch = -1; end  % if waveform outside bounds,
25.         if x(1) < 0.0001, latch = 1; end   % reset latch as appropriate to
26.         tout = output(t,x_var,y_var);  % change direction of integration
27.     end                                % output results at end of step
```

A.2.14   SAWTEVIN.M

Copyright © B S Bennett, 1994

```
1.  % This is a SAWTOOTH WAVEFORM GENERATOR, producing the same
2.  % waveform as before, but  with event processing
3.  % to deal with discontinuities

4.  % Declare all global variables required in the simulation
5.      global h_min iter_ate t_out x_out y_out
6.      global run_plot xy_plot xy_size abs_xy a_a b_b x_label
7.      global points t_last y_last x_last v_plot
8.      global x_var y_var e_vent
9.  % Set run conditions
10.     t0 = 0;                            % simulation start time
11.     run_plot = 1;                      % enable run-time plotting
```

```
12.    tfin = 6;                          % simulation finish time
13.    x0 = 0.0;                          % initial condition
14.    cint = (tfin - t0)/60;             % output interval
15.    h_min = cint/1000;                 % minimum allowable step size
16.    v_plot = [t0,tfin,0,1;t0,tfin,-2,4];  % run-time plot scaling
17.    xy_plot = ['x1';'y1'];             % run-time plot variable names
18.    x_label = 'time';                  % Independent variable name
19.    [dum1] = decode(t0);               % run-time plot parameters

20. % Start the simulation
21.    [t,x] = feval('dynamb','ode23b','sawtevb',t0,tfin,x0,cint);

22. % Return graphics screen to default state and produce post-run plots
23.    shg,pause,hold off
24.    subplot
25.    plot(t_out,x_out,t_out,y_out(:,1))
26.      title('sawtooth waveform with event processing'),pause
27.    plot(t_out,x_out,'og')
28.      title('sawtooth waveform with event processing'),pause
29.     plot(t_out,x_out,'.g')
30.      title('sawtooth waveform with event processing'),pause
31.    clg
```

A.2.15 SAWTEVB.M

Copyright © B S Bennett, 1994

```
1. % This is the model which generates a SAWTOOTH WAVEFORM with the aid of
2. % event processing.  By setting event times to coincide with the peaks
3. % and troughs of the waveform, integration step size can be adjusted to
4. % coincide with these events, and the discontinuities brought about by
5. % slope reversal are thereby negotiated with ease.

6. function dx = sawtevb(t,x,control)

7.    if control == 1                    % INITIAL REGION
8.        y_var(1) = 0;                  % initialise a y-variable
9.    end                                % for use as waveform slope


10.    if control == 2                   % DYNAMIC REGION.
11.        dx(1) = y_var(1);             % calculate waveform derivative
12.    end


13.    if control == 5                   % PROCESS EVENT OF CLASS 1
14.        if y_var(1) > 0               % if slope is positive, THEN
15.            e_vent(1) = t + 1.0/1.5;  % set next event, based on negative
16.            y_var(1) = -1.5;          % slope. Set slope negative.
17.        else                          % OTHERWISE
18.            e_vent(1) = t + 1.0/3.0;  % set next event, based on positive
19.            y_var(1) = 3.0;           % slope. Set slope positive.
20.        end
21.    end
```

Appendix A.3

MATSIM Simulation Routines

The files listed in this appendix are M-files that have collectively been given the name MATSIM. MATSIM is a set of four routines that run in association with the pairs of files as listed in Appendix A.2. The process is set in motion by typing (against the MATLAB prompt) the name of the initialization file of the pair (*IN.M). This calls DECODE.M if required for run-time plotting, followed by DYNAMB.M which is the main run-time executive. This in turn calls ODE23B.M and OUTPUT.M to generate the solution to the model equations and, if enabled, a run-time plot of one or more solution transients. The model file of the pair (*B.M) is called repeatedly by ODE23B.M, and on occasion by DYNAMB.M.

If you are interested in obtaining a display of certain model or system variables as the solution proceeds, simply remove the semi-colon at the end of the statement line where that variable is evaluated.

A.3.1 DYNAMB.M

Copyright © B S Bennett, 1994

```
1.  % This is the main control routine for running a simulation.   It includes
2.  % elements for obtaining solution output at designated output event points, and
3.  % before and after processing model events.   It includes event-processing
4.  % of up to six different types of model event for handling discontinuities,
5.  % and for the simulation of computer control.   It allows for single-run
6.  % solution (normal case) or iterative multiple-run solution
7.  % (obtained by setting [iter_ate=1] repeatedly in the model terminal region
8.  % until specified criteria are satisfied).
9.  function [tout,xout] = dynamb(algorithm,eqns,t0,tfin,x0,cint)
10. control = 1:10;        % set up a program flow control vector
11. e_vent = zeros(1,6);   % set up an event vector for event-processing
12. hzero = h_min/10;      % set minimum time-step assumed non-zero
13. iter_ate = 1;          % set flag positive for iterative operation
14.            % this is the beginning of the run iteration sequence
15. while iter_ate == 1    % repeat simulation run while flag positive
16.    print = t0;         % initialise output time to time zero
```

```
17.    t_out = [];        % initialise vector for time output storage
18.    x_out = [];        % initialise vector for x-vector output storage
19.    y_out = [];        % initialise vector for y-vector output storage
20.    points = 0;        % reset to zero number of output points stored
21.    rand('seed',0)     % reset random number generator (in case it is used)
22.    t_yy = [];         % initialise time vector for noise function storage
23.    y_yy = [];         % initialise amplitude vector for noise function storage
24.    t = t0;            % initialise simulation time
25.    x_var = x0;        % initialise run-time x-vector
26.    dx = feval(eqns,t,x0,control(1));  % set up initial conditions for run
27.    dx = feval(eqns,t,x0,control(2));  % evaluate derivatives at time zero
28.    loop = 1;  % enable operation of infinite loop for stepping through
29.              % solution to equation
30.    while loop == 1   % commence solution
31.         % if output time or end of run reached, activate output
32.         if t == print | t >= tfin
33.             dx = feval(eqns,t,x_var,control(2));  % update model equations
34.             tout = output(t,x_var,y_var);          % output current values
35.             print = print + cint;                  % set next output time
36.         end
37.         % event-processing section: process each event that coincides
38.         % with the current simulation time
39.         for i=1:6
40.             if abs(t-e_vent(i))<hzero             % if event reached
41.                 dx = feval(eqns,t,x_var,control(2));   % update model equations
42.                 tout = output(t,x_var,y_var);          % output pre-event values
43.                 dx = feval(eqns,t,x_var,control(4+i)); % process the event
44.                 dx = feval(eqns,t,x_var,control(2));   % update model equations
45.                 tout = output(t,x_var,y_var);          % output post-event values
46.             end
47.         end
48.         if t>=tfin, break, end;             % if run time reached stop solution
49.         if isempty(e_vent(find(e_vent>0)));  % set length of next integration step
50.             hin = min(print-t, tfin-t);
51.         else
52.             hin = min(min(print-t, tfin-t), min(e_vent(find(e_vent>0))-t));
53.         end
54.         tin = t;
55.         [t,x] = feval(algorithm,eqns,tin,hin);   % integrate over this time step
56.     end
57.     iter_ate = 0;                          % disable solution loop iteration
58.     dx = feval(eqns,t,x_var,control(2));   % update model equations
59.     tout = output(t,x_var,y_var);          % output solution values
60.     dx = feval(eqns,t,x_var,control(3));   % carry out any post-run (terminal region)
61.                                            % processing that is required.
62.         % This includes resetting iter_ate to 1 for a repeat run, if needed.
63. end
```

A.3.2 DECODE.M

```
1.  % Run-time plotting is carried out by assigning each variable to be plotted
2.  % (up to a maximum of 4) to a sub-plot set of axes.   The number and
```

```
3.  % orientation of the sub-plots to be produced is determined by this
4.  % routine.   The two arrays returned are
5.  %     (a) the subplot orientation parameters and
6.  %     (b) the indices of the x- and/or y-variables to be plotted.


7.  function [dum1] = decode(dum2)


8.    xy_size = size(xy_plot);  % number and sizes of plot variable name strings
9.    abs_xy = abs(xy_plot);    % ASCII values for name strings


10.   for i = 1:xy_size(1)                   % For each run-time plot variable
11.       a_a(i) = 200+10*fix(xy_size(1)/3+1)+i;  %  determine subplot orientation
12.       if xy_size(2)==2                    %  parameter, and indices of
13.           b_b(i) = abs_xy(i,2)-48;        %  run-time variables to be
14.       else                                %  plotted.
15.           if abs_xy(i,3)==32
16.               b_b(i) = abs_xy(i,2)-48;
17.           else
18.               b_b(i) = 10*(abs_xy(i,2)-48) + abs_xy(i,3)-48;
19.           end
20.       end
21.   end
```

A.3.3 ODE23B.M

Copyright © B S Bennett, 1994

```
1.  % This routine is called in an attempt to integrate the model differential
2.  % equations from an initial time [tin], through a desired solution interval
3.  % [hin], whose value is determined by the calling routine DYNAM, in terms of
4.  % output or process events.   The simulation time at the end of the desired
5.  % solution interval [tout] is not necessarily reached in one integration step.
6.  % The integration routine is applied iteratively until this time is reached.
7.  % The routine used is based on the following reference:
8.  % NIESSE, D H : Technical comment on low-order Runge-Kutta variable step
9.  %                 integration methods
10. %                 SIMULATION, Vol. 14, No. 2, (February, 1970), pp. 93-94


11. function [tout,xout] = ode23b(eqns,tin,hin)


12. t = tin;           % set time at beginning of integration interval
13. tol = 1.0e-3;      % set allowable error tolerance
14. tout = tin + hin;  % set time at end of integration interval
15. h = hin;           % set initial step size to desired integration interval
16. order = 1/3;       % set order of integration method
17. while (t<=tout)&(h>=h_min)  % repeat integration until [tout] is reached
18.                             % or integration step size too small
19.     xlast = x_var;   % save x-vector values at time [tin]
20.     control = 2;     % set for solution of the model dynamic equations
21.     dx1 = feval(eqns,t,xlast,control);              % 1st stage of RK integration
22.     dx2 = feval(eqns,t+h/2,xlast+0.5*h*dx1,control);  % 2nd stage
23.     dx3 = feval(eqns,t+h,xlast-h*(dx1-2*dx2),control); % 3rd stage
24.     err = norm(h*(dx1-2*dx2+dx3)/3,'inf');  % calculate error estimate
```

```
25.     errmax = tol*max(norm(xlast,'inf'),1.0); % calculate permissible error
26.     if err<=errmax              % if error criterion is passed, then
27.         t = t + h;                      % update simulation time
28.         x_var = xlast + h*(dx1+4*dx2+dx3)/6; % update solution x-vector
29.         control = 4;    % set flow control to carry out end-of-step processing
30.         xin = x_var;    % save updated solution vector
31.         dx = feval(eqns,t,xin,control); % carry out end-of-step processing
32.     else
33.         x_var = xlast;  % reset x-vector to values at beginning of step
34.     end
35.     if err ~= 0.0                           % if error estimate is non-zero,
36.                                             % adjust next integration step
37.         h = min(0.9*h*(errmax/err)^order, tout-t); % in terms of error ratio
38.     else            % otherwise adjust in terms of remainder of desired
39.         h = tout-t;     % integration interval remaining for this call to ODE
40.     end
41. end
42. xout = x_var;  % update output x-vector
```

A.3.4 OUTPUT.M

Copyright © B S Bennett, 1994

```
1.  % When it is called at every designated output time during the simulation,
2.  % both at normal regularly spaced output points and at other event times,
3.  % this routine
4.  %      (a) stores current values for the output variables in a number of
5.  %          arrays for post-run plotting, and if run plotting is enabled,
6.  %      (b) Draws one or more sets of run-time plot axes (at the beginning of
7.  %          the simulation)
8.  %      (c) Produces (by vector increments) the appropriate run-time plots.


9.  function tout = output(t,x,y)


10.     t_out = [t_out;t];      % add current value of time to time output array
11.     x_out = [x_out;x_var];  % add current x-values to output x-array
12.     y_out = [y_out;y_var];  % add current y-values to output y-array


13.     if run_plot == 1        % If run-time plotting is enabled, then
14.         if points==0        %       on the first call only,
15.             t_last = t;      %       save initial values.
16.             x_last = x_var;  %       Do nothing more.
17.             y_last = y_var;
18.         else                % Otherwise form (* x 2) arrays for
19.             t2 = [t_last;t];        % simulation time ) incorporating values
20.             x2 = [x_last;x_var];    % x-values       ) from the last output
21.             y2 = [y_last;y_var];    % y-values       ) point to the present
22.             if isempty(xy_plot) == 0
23.                 for i = 1:xy_size(1)    % For each run-time plot to be produced
24.                     axis(v_plot(i,:));  %   scale the axes
25.                     if points == 1      % At time zero
26.                         subplot(a_a(i)); %   draw and label axes for plots
```

```
27.                      ylabel(xy_plot(i,:));
28.                       xlabel(x_label);
29.                        title('run-time plot');
30.                    end                      % At all output times
31.                    subplot(a_a(i));         % Select the appropriate sub-plot
32.                    if abs_xy(i,1)==120      % and add an incremental vector to
33.                       plot(t2,x2(:,b_b(i))) % the run-time transient being
34.                    else                     % produced within that set of axes.
35.                        plot(t2,y2(:,b_b(i)))
36.                    end
37.                end
38.               hold on    % Set flag to retain plots, and add increments on
39.            end            % each subsequent call.
40.        end
41.        t_last = t;            % save current value of time
42.        x_last = x_var;        % save current x-values
43.        y_last = y_var;        % save current y-values
44.        points = points + 1;   % increment number of points plotted/stored
45.    end
```

Index